Functional Importance of the Plant Microbiome

Sharon Lafferty Doty
Editor

Functional Importance of the Plant Microbiome

Implications for Agriculture, Forestry and Bioenergy

Springer

Editor
Sharon Lafferty Doty
School of Environmental and Forest Sciences
College of the Environment
University of Washington
Seattle, WA, USA

ISBN 978-3-319-65896-4 ISBN 978-3-319-65897-1 (eBook)
DOI 10.1007/978-3-319-65897-1

Library of Congress Control Number: 2017953942

Printed on acid-free paper

This Springer imprint is published by Springer Nature
The registered company is Springer International Publishing AG
The registered company address is: Gewerbestrasse 11, 6330 Cham, Switzerland

Contents

Chapter 1
Functional Importance of the Plant Endophytic Microbiome: Implications for Agriculture, Forestry, and Bioenergy

Sharon Lafferty Doty

Just as the human microbiome is important for our health [1], so too the plant microbiome is necessary for plant health, but perhaps more so. Since plants cannot move, they face more challenges in acquiring sufficient nutrients from a given site, defending against herbivores and pathogens, and tolerating abiotic stresses including drought, salinity, and pollutants. The plant microbiome may help plants overcome these challenges. Since genetic adaptation is relatively slow in plants, there is a distinct advantage to acquiring an effective microbiome able to more rapidly adapt to a changing environment. Although rhizospheric microorganisms have been extensively studied for decades, the more intimate associations of plants with endophytes, the microorganisms living fully within plants, have been only recently studied. It is now clear, though, that the plant microbiome can have profound impacts on plant growth and health. Comprising an ecosystem within plants, endophytes are involved in nutrient acquisition and cycling, interacting with each other in complex ways. The specific members of the microbiome can vary depending on the environment, plant genotype, and abiotic or biotic stresses [2–6]. The microbiome is so integral to plant survival that the microorganisms within plants can explain as much or more of the phenotypic variation as the plant genotype [7]. In plant biology research, an individual plant should thus be viewed as a whole, the plant along with intimately associated microbiota (a "holobiont"), with the microbiome playing a fundamental role in the adaptation of the plant to environmental challenges [8–10].

Intensive agriculture has stripped away many of the natural partnerships that plants in their native environments would have depended upon. Consequently, the services once provided through symbiosis have been replaced with chemical fertilizers, pesticides, and other inputs. However, as the functional significance of the

S.L. Doty (✉)
School of Environmental and Forest Sciences, College of the Environment,
University of Washington, Seattle, WA 98195-2100, USA
e-mail: sldoty@uw.edu

S.L. Doty (ed.), *Functional Importance of the Plant Microbiome*,
DOI 10.1007/978-3-319-65897-1_1

microbiome has been revealed [11], the importance of restoring these relationships to optimize plant growth and yields in an environmentally sustainable manner has been recognized [12]. Plant microbiome sequencing has been performed for several plant species which is an important first step; however, the resulting data often provide only the species-level organization, not the necessary functional clues gained from metagenomic sequencing. For example, a metagenomic analysis of the full endophyte community of roots of field-grown rice provided extensive information on genes associated with the endophytic lifestyle [13]. Genes involved in N-fixation, phytohormone production, ROS detoxification, ACC deaminase, transport systems, signaling, colonization, and other putative symbiosis-related genes were identified, providing information on the functional attributes of the plant microbiome [13]. As sequencing technologies progress, it will be increasingly more feasible to gather this crucial information [14], more crucial since choice of the members of the plant microbiome is likely to be microbial strain specific, not species specific, with plants selecting for particular attributes important for the particular environmental condition [15, 16].

Through a better understanding of beneficial plant-microbiome interactions, improvements in the economic and environmental sustainability of agriculture, forestry, and bioenergy can be achieved. In all three of these industries, a reduction in inputs, whether it be fertilizer, water, or chemical pesticides, would lead to significant cost savings. The cost of using N-fixing microbes is estimated to be only 1% as much as that of using chemical fertilizers [17]. Chemical N fertilizers are produced using fossil fuels, high temperatures, and high pressure. Since diazotrophic endophytes use plant sugars produced from solar energy, the dinitrogen gas abundant in the atmosphere is fixed by the bacteria, providing essentially continuous fertilizer for the crops at little financial cost. In addition to nitrogen, the other main component of chemical fertilizer is phosphate. Since the majority of phosphate in soils is inaccessible to plants, there is rising demand, and there is a finite supply of rock phosphate; the cost of this key macronutrient is rising. Seedling mortality due to drought also results in major financial losses. With the increased frequency and duration of drought, the cost of freshwater rights can become a determining factor in deciding which crops to grow, as it incentivizes the cultivation of only the highest value crops to make up for the high cost of water. Specific endophyte strains can defend the host plant against pathogens [18, 19], potentially reducing the need for chemical pesticides. Through increasing plant growth and crop yields, and decreasing the amounts of inputs including fertilizers, water, and pesticides, endophytes have the potential to increase profit margins. The use of bio-stimulants has recently gained popularity among agricultural biotechnology companies, with the global market for bio-stimulants for plants estimated to rise to USD 3.6 billion by 2022 [20]. In addition to the economic benefits of appropriate endophyte inoculations, substantial improvements in the environmental sustainability of these industries can be made by lessening the impacts to aquatic ecosystems from chemical run-off, reducing greenhouse gas emissions, and lowering the depletion rate of groundwater reserves.

While the positive implications of endophyte inoculations for agriculture have been well reviewed [20–25], less attention has been given to potential impacts on forestry [26, 27]. Successful inoculation of widely used conifer species with endophytes could increase forest productivity and reduce reforestation costs, particularly through reducing tending costs during early stand management. Major advantages of using natural plant-microbe symbioses are that (1) they are easily applied at the greenhouse stage prior to out-planting, (2) the increased drought tolerance can occur just weeks after inoculation, (3) the microorganisms are easy and inexpensive to grow, and (4) they provide multiple benefits including increased nutrient acquisition, drought tolerance, growth, and overall health. By augmenting the microbiome of nursery stock at the greenhouse stage, foresters and restoration practitioners may be able to reduce the mortality rate during establishment. With the increased frequency and duration of drought, and the increased cost of fertilizers, the improved resilience and growth of trees from bio-inoculants would be an economic advantage for the forestry industry.

With limited arable lands and resources for both agriculture and bioenergy production, biomass for bioenergy should ideally be produced with fewer inputs and on marginal lands without competing with agriculture. Symbiosis with microorganisms can allow plants to overcome the challenges faced in these environments, including low-nutrient soils with limited water. Overcoming such challenges will be even more crucial when they are confronted with the increased temperatures and re-localization of precipitation seen with climate change. By understanding the natural plant-microbe interactions at work to increase plant stress tolerance in biomass crops, symbiosis-based technologies may be developed to increase biomass production. *Populus* is a flagship genus for the production of environmentally sustainable biomass for cellulosic ethanol and biochemicals. Endophytes from hybrid poplar have also been shown to increase growth of this important bioenergy crop [28–30]. Endophytes from native poplar applied to hybrid poplar can increase photosynthetic efficiency [31], drought tolerance [32], N_2-fixation [33], and a doubling of root mass accumulation [33]. This increased rooting could improve below-ground carbon storage and may also help with drought tolerance. Plants can increase photosynthetic rates under elevated CO_2 conditions only until other factors such as nutrients and water become limiting. Diazotrophic endophytes that also increase drought tolerance could therefore be used to improve the growth and sustainability of biomass production.

The focus of this book is on the functional importance of endophytes to plant growth and health. Endophytes can increase nutrient availability for plants through nitrogen fixation (Chap. 2), phosphate solubilization, and siderophore production (Chap. 3). The phytobiome can improve photosynthetic efficiency and water use efficiency (Chap. 4). Specific endophytes can increase tolerance to abiotic stresses including temperature, drought, and salinity (Chap. 5). Many endophyte strains are capable of producing hormones or modulating the host phytohormones, improving both plant growth and stress tolerance (Chap. 6).

To maximize the benefits of these symbioses, further research is required to understand at the mechanistic level how endophytes perform all of these integral roles for the host plant. It is time for a greener revolution, not based on chemical applications but on natural plant-microbe partnerships.

References

1. Gordon JI. Honor thy gut symbionts redux. Science. 2012;336:1251–3.
2. Bonito G, Reynolds H, Robeson MS, Nelson J, Hodkinson BP, Tuskan G, Schadt CW, Vilgalys R. Plant host and soil origin influence fungal and bacterial assemblages in the roots of woody plants. Mol Ecol. 2014;23:3356–70.
3. Shakya M, Gottel N, Castro H, Yang ZK, Gunter L, Labbe J, Muchero W, Bonito G, Vilgalys R, Tuskan G, Podar M, Schadt CW. A multifactor analysis of fungal and bacterial community structure in the root microbiome of mature Populus deltoides trees. PLoS One. 2013;8:e76382.
4. Lundberg DS, Lebeis SL, Paredes SH, Yourstone S, Gehring J, Malfatti S, Tremblay J, Engelbrektson A, Kunin V, del Rio TG, Edgar RC, Eickhorst T, Ley RE, Hugenholtz P, Tringe SG, Dangl JL. Defining the core Arabidopsis thaliana root microbiome. Nature. 2012;488:86–90.
5. Gottel NR, Castro HF, Kerley M, Yang Z, Pelletier DA, Podar M, Karpinets T, Uberbacher E, Tuskan GA, Vilgalys R, Doktycz MJ, Schadt CW. Distinct microbial communities within the endosphere and rhizosphere of Populus deltoides roots across contrasting soil types. Appl Environ Microbiol. 2011;77:5934–44.
6. Edwards J, Johnson C, Santos-Medellin C, Lurie E, Podishetty NK, Bhatnagar S, Eisen JA, Sundaresan V. Structure, variation, and assembly of the root-associated microbiomes of rice. Proc Natl Acad Sci U S A. 2015;112:E911–20.
7. Friesen ML, Porter SS, Stark SC, von Wettberg EJ, Sachs JL, Martinez-Romero E. Microbially mediated plant functional traits. Annu Rev Ecol Evol Syst. 2011;42:23–46.
8. Vandenkoornhuyse P, Quaiser A, Duhamel M, Le VA, Dufresne A. The importance of the microbiome of the plant holobiont. New Phytol. 2015;206:1196–206.
9. Hartmann A, Rothballer M, Hense BA, Schroder P. Bacterial quorum sensing compounds are important modulators of microbe-plant interactions. Front Plant Sci. 2014;5:131.
10. Rodriguez RJ, Redman RS. Moret than 400 million years of evolution and some plants still can't make it on their own: plant stress tolerance via fungal symbiosis. J Exp Bot. 2008;59:1109–14.
11. Bulgarelli D, Schlaeppi K, Spaepen S, Ver Loren van Themaat E, Schulze-Lefert P. Structure and functions of the bacterial microbiota of plants. Annu Rev Plant Biol. 2013;64:807–38.
12. Busby PE, Soman C, Wagner MR, Friesen ML, Kremer J, Bennett A, Morsy M, Eisen JA, Leach JE, Dangl JL. Research priorities for harnessing plant microbiomes in sustainable agriculture. PLoS Biol. 2017;15:e2001793.
13. Sessitsch A, Hardoim P, Doring J, Weilharter A, Krause A, Woyke T, Mitter B, Hauberg-Lotte L, Friedrich F, Rahalkar M, Hurek T, Sarkar A, Bodrossy L, van Overbeek L, Brar D, van Elsas JD, Reinhold-Hurek B. Functional characteristics of an endophyte community colonizing rice roots as revealed by metagenomic analysis. Mol Plant-Microbe Interact. 2012;25:28–36.
14. Burton JN, Liachko I, Dunham MJ, Shendure J. Species-level deconvolution of metagenome assemblies with Hi-C-based contact probability maps. G3 (Bethesda). 2014;4:1339–46.
15. Burke C, Steinberg P, Rusch D, Kjelleberg S, Thomas T. Bacterial community assembly based on functional genes rather than species. Proc Natl Acad Sci U S A. 2011;108:14288–93.
16. Reinhold-Hurek B, Bunger W, Burbano CS, Sabale M, Hurek T. Roots shaping their microbiome: global hotspots for microbial activity. Annu Rev Phytopathol. 2015;53:403–24.
17. Stokstad E. The nitrogen fix. Science. 2016;353:1225–7.
18. Rybakova D, Cernava T, Koberl M, Liebminger S, Etemadi M, Berg G. Endophytes-assisted biocontrol: novel insights in ecology and the mode of action of Paenibacillus. Plant Soil. 2016;405:125–40.
19. Busby PE, Ridout M, Newcombe G. Fungal endophytes: modifiers of plant disease. Plant Mol Biol. 2016;90:645–55.
20. Timmusk S, Behers L, Muthoni J, Muraya A, Aronsson AC. Perspectives and challenges of microbial application for crop improvement. Front Plant Sci. 2017;8:49.

21. Compant S, Clement C, Sessitsch A. Plant growth-promoting bacteria in the rhizo- and endosphere of plants: their role, colonization, mechanisms involved and prospects for utilization. Soil Biol Biochem. 2010;42:669–78.
22. Mahanty T, Bhattacharjee S, Goswami M, Bhattacharyya P, Das B, Ghosh A, Tribedi P. Biofertilizers: a potential approach for sustainable agriculture development. Environ Sci Pollut Res Int. 2017;24:3315–35.
23. Lugtenberg BJ, Caradus JR, Johnson LJ. Fungal endophytes for sustainable crop production. FEMS Microbiol Ecol. 2016;92
24. Le CK, Gurr SJ, Hirsch PR, Mauchline TH. Exploitation of endophytes for sustainable agricultural intensification. Mol Plant Pathol. 2017;18:469–73.
25. Bhardwaj D, Ansari MW, Sahoo RK, Tuteja N. Biofertilizers function as key player in sustainable agriculture by improving soil fertility, plant tolerance and crop productivity. Microb Cell Factories. 2014;13:66.
26. Germaine KJ, Chhabra S, Song B, Brazil D, Dowling DN. Microbes and sustainable production of biofuel crops: a nitrogen perspective. Biofuels. 2010;1:877–88.
27. Santoyo G, Moreno-Hagelsieb G, Del CO-M, Glick BR. Plant growth-promoting bacterial endophytes. Microbiol Res. 2016;183:92–9.
28. Hacquard S, Schadt CW. Towards a holistic understanding of the beneficial interactions across the Populus microbiome. New Phytol. 2015;205:1424–30.
29. Scherling C, Ulrich K, Ewald D, Weckwerth W. A metabolic signature of the beneficial interaction of the endophyte paenibacillus sp. isolate and in vitro-grown poplar plants revealed by metabolomics. Mol Plant Microbe Interact. 2009;22:1032–7.
30. Weyens N, van der Lelie D, Taghavi S, Newman L, Vangronsveld J. Exploiting plant-microbe partnerships to improve biomass production and remediation. Trends Biotechnol. 2009;27:591–8.
31. Knoth J, Kim S-H, Ettl G, Doty SL. Effects of cross host species inoculation of nitrogen-fixing endophytes on growth and leaf physiology of maize. GCB Bioenergy. 2013;5:408–18.
32. Khan Z, Rho H, Firrincieli A, Luna V, Hung SH, Kim S-H, Doty SL. Growth enhancement and drought tolerance of hybrid poplar upon inoculation with endophyte consortia. Curr Plant Biol. 2016;6:38–47.
33. Knoth JL, Kim SH, Ettl GJ, Doty SL. Biological nitrogen fixation and biomass accumulation within poplar clones as a result of inoculations with diazotrophic endophyte consortia. New Phytol. 2014;201:599–609.

Chapter 2
Endophytic N-Fixation: Controversy and a Path Forward

Sharon Lafferty Doty

Nitrogen (N) is an essential macronutrient due to its being a component of proteins, nucleic acids, and the energy currency of cells, ATP. While nearly 80% of our atmosphere is comprised of N, it is in an inert form, inaccessible to most life forms. Lightning strikes convert N_2 gas into ammonia or nitrate that is deposited in soils through rainfall, accounting for approximately 10% of available N [1]. Biological reduction of the triple bond of N_2 gas, however, requires the nitrogenase complex, a multi-subunit enzyme found in a subset of archaeal and bacterial species. This oxygen-sensitive complex "fixes" the dinitrogen gas into ammonia through an energy-intensive process, requiring 20–30 molecules of ATP per molecule of dinitrogen gas under normal physiological conditions [2]. Plants acquire N from soils rich in organic matter where previously fixed N is made available through decomposition, but where soils are nutrient poor, N is the key nutrient limiting growth.

The so-called green revolution of the twentieth century was made possible through the Haber-Bosch process for production of chemical N fertilizer. Using high temperature (400–650° C) and pressure (200–400 atm), and approximately 2% of global fossil fuels, this method produces over 450M tons of N fertilizer each year [3]. While this process is effective, its widespread use in commercial agriculture is not environmentally sustainable. Levels of ammonia in the atmosphere have increased significantly as a result of intensive agricultural practices [4]. Only about half of the applied fertilizer is taken up by plants. The excess N is converted by soil microorganisms to nitrous oxide, a potent greenhouse gas, or is leached into aquatic systems, disrupting the natural ecosystems [5].

Another source of fixed N for plants relies on biological N-fixation. Select groups of plants have evolved intimate partnerships with N-fixing (diazotrophic) bacteria harbored in specialized organs, termed nodules, most commonly found on the roots

S.L. Doty (✉)
School of Environmental and Forest Sciences, College of the Environment,
University of Washington, Seattle, WA 98195-2100, USA
e-mail: sldoty@uw.edu

S.L. Doty (ed.), *Functional Importance of the Plant Microbiome*,
DOI 10.1007/978-3-319-65897-1_2

of such plants. These nodulated plants include the legumes and *Parasponia* that associate with rhizobia and the actinorhizal plants that associate with *Frankia*. The symbiotic interactions between the N-fixing bacteria and host plants have been extensively studied with a recent focus on the signaling mechanisms that trigger the specific association [6, 7]. Following mutual recognition, a root nodule forms, providing a specialized organ for symbiotic N-fixation in which photosynthate is exchanged for fixed N [8, 9]. Root nodules are generally 2–5 mm in diameter and are occupied by up to 10^9 rhizobia [1]. Since the nitrogenase complex is oxygen labile [10] and yet requires high levels of ATP for the reaction, legumes express a leghemoglobin that maintains a low-oxygen environment in the nodule, while the diazotrophic bacteria use a high-oxygen-affinity cytochrome oxidase for oxidative phosphorylation. The nodule further limits oxygen with an oxygen-diffusion barrier. While providing an apparently ideal environment for N-fixation, legumes tightly control the housed bacteria, sanctioning nutrients to favor the most effective symbionts [11].

There are many natural environments in which organic N is limiting and yet non-nodulating plant species thrive. From where do these plants obtain this essential nutrient? Over the last few decades, studies have demonstrated that N-fixing bacteria can be found throughout the plant body of such plants, tightly bound to the plant surface or within the plant in the apoplastic intercellular spaces or within plant cells. So-called associative and endophytic diazotrophic bacteria (AEDB), these bacteria may be specifically recognized by the host plant [12]. Unlike rhizobia that commonly use an infection thread to enter the plant host, endophytes use crack-entry, colonizing the lateral root junctions, and migrating within the plant. N-fixing endophytic bacteria were first isolated from grasses such as kallar grass [13], sugarcane [14], wild rice [15], and maize [16, 17], but also from a wide variety of plant species including African sweet potato [18], rock-colonizing cactus [19], miscanthus [20], feather mosses [21], dune grasses [22], coffee plants [23], invasive grasses [24], and poplar and willow [25–27]. Significant rates of biological N-fixation (BNF) from AEDB have been recorded [28–34]. Through isolations of culturable endophytic strains from the native hosts and re-inoculation into host or non-host plant species, it has been demonstrated in multiple studies over the last few decades that plants often benefit from these endophytic microorganisms with increased health and growth. Significant N-fixation has been quantified in some of these cases, such as in sugarcane [35], wheat [36], rice [37], lodgepole pine [38], and Western redcedar [39].

Despite several decades of global research on N-fixation in a diversity of non-nodulated plants, it is a widely held belief that only symbiotic N-fixation in root nodules is significant for plant growth. This view has led to the recent focus on systems biology transgenic approaches to attempt to solve the global fertilizer problem [3]. One approach is to engineer non-legumes to express a functional nitrogenase complex [40] while another approach is to engineer them to form root nodules in which rhizobia would fix N [41]. While both approaches are extraordinarily complex and would be major scientific achievements, they suffer from two basic problems: they are not widely applicable to crops other than the specific, engineered lines, and many countries restrict the use of transgenic crops.

Alternatively, diazotrophic endophytes have a broad host range and public acceptance, and can be commercially viable. Unlike the limitation of rhizobia for legumes, some endophyte strains can increase growth of many different plant species. For example, diazotrophic endophytes from Salicaceae species colonize and improve growth of other eudicots [42], monocots [27, 43, 44], and even conifers [45]. With an understanding of the importance of the human microbiome and the advent of probiotics, the public seems more willing to accept bio-inoculants than genetically modified crops. Biofertilizers, living products that promote plant growth by increasing nutrient supply, provide a cost-effective method to elevate crop yields and improve soil fertility [31, 46]. The biofertilizer market has increased dramatically, and is expected to reach USD 1.66 billion by 2022 but does have challenges [47]. Economic advantages for farmers are that treating a field with microbial inoculum costs just 1% as much as adding N fertilizer [3] and may require only one application per season if the strains are compatible. Biofertilizers can be applied as peat formulations, liquid formulations, granules, and freeze-dried powders [46], or applied to the seeds directly prior to planting. An environmental advantage of using endophytes is that, since N is fixed within the plants, it is less likely to have the issues associated with excess N fertilizer in soils.

Despite decades of research and enormous potential for increasing plant growth with reduced inputs [46], endophytic N-fixation is still portrayed as being not possible or as insignificant. This review summarizes the arguments against associative N-fixation, poses explanations in support of the importance of endophytic diazotrophs, and suggests future directions for research.

2.1 The Oxygen Issue

The most common argument made against endophytic N-fixation is the misconception that only root nodules have the required low-oxygen environment conducive to N-fixation. The nitrogenase enzyme is exquisitely sensitive to oxygen [48]. Paradoxically, the N-fixation reaction requires such high levels of ATP that it is advantageous to have oxidative metabolism to provide sufficient energy to drive the reaction [2]. In legumes, the high demand for oxygen in root nodules is supplied by leghemoglobin [49], and legume nodules have a peripheral vasculature that would supply oxygen for generating high ATP levels [1]. Free-living, aerobic N-fixing bacteria evolved multiple methods to protect nitrogenase from oxygen, and microaerobic environments conducive to N-fixation do exist within plant tissues. These arguments are discussed below.

Nitrogenase is found in a variety of free-living microorganisms including obligate aerobes and oxygenic phototrophs. Phylogenetic evidence indicates that nitrogenase evolved first in anaerobic archaea, spreading into aerobic bacteria with additional sets of genes not for direct protection of the enzyme from oxygen but rather for regulation of gene expression and optimization for the increased N demands of aerobic growth [50]. Of the factors that correlated with the increased complexity of the nitrogenase-related genes between the archaea and bacteria, metabolic protection of the nitrogenase enzyme was the only factor known to be involved in protection from oxygen [50]. High respiration rates rapidly consume oxygen during uncoupled respiration [51]. Other common methods to protect nitrogenase from oxygen include temporal protection (fixing N at night when oxygen tensions are lower) and spatial protection by compartmentalization of the enzyme [2, 52]. While some photosynthetic, N-fixing cyanobacteria have thick-walled specialized cells, termed heterocysts, in which N-fixation takes place, more N is fixed in some marine environments by non-heterocyst cyanobacteria [53, 54]. For example, *Trichodesmium*, an ancient non-heterocystous marine diazotroph, uses temporal regulation, fixing N when oxygen consumption exceeds photosynthetic oxygen production [55]. A well-studied aerobic diazotrophic soil bacteria is *Azotobacter vinelandii* that uses multiple mechanisms to protect nitrogenase including respiratory, conformational, and autoprotection mechanisms [2, 51, 56]. High respiration rates are essential for aerobic N-fixation in *A. vinelandii* [2]. The genome sequence of *A. vinelandii* revealed genes for multiple terminal oxidases for reducing cellular oxygen levels, and for the alginate capsule that may impede oxygen transfer into the cell [56]. It is clear that aerobic bacteria are capable of preventing inhibition of nitrogenase by oxygen.

A second argument supporting the possibility of endophytic N-fixation is that microaerobic environments likely do exist within plants without root nodules. Tissues of stem nodes are dense and would have a low oxygen level [57]. The apoplast and vascular tissues are low in oxygen and could provide appropriate conditions [9, 57, 58]. Endophytes tend to form microcolonies within plants, and since biofilms have limited oxygen within the interior [59], these aggregates could be conducive to N-fixation. For example, the sugarcane endophyte, *Gluconacetobacter diazotrophicus* forms mucilaginous microcolonies in the apoplast, and it was shown in vitro that the mucilage provided resistance to oxygen diffusion, providing sufficient oxygen for bacterial respiration without inhibition of nitrogenase [60, 61]. Nitrogenase gene expression is regulated by oxygen [62, 63]. Several studies using *nif* fusions with reporters or nitrogenase antibodies have demonstrated that endophytes do express the genes in planta. For example, the *nifH* genes of *Azoarcus* were expressed in Kallar grass roots [64] and in rice [65]. In maize, co-localization of GFP-tagged *Klebsiella pneumoniae* with nitrogenase immunolocalization showed that nitrogenase is expressed by the cells; however, it required exogenous carbon source in this study [66]. In wheat, *K. pneumoniae* expressed nitrogenase protein in the intercellular space of the root cortex [36]. On the root surface of *Setaria viridis*, *nifH*-gus fusions showed that the nitrogenase gene is expressed by *Azospirillum brasilense* [67]. Diazotrophic *Herbaspirillum* sp. expressed nitrogenase while epiphytic or endophytic with rice, sorghum, wheat, and maize (reviewed in [68]). A recent RNA-seq transcriptional profiling study of *H. seropedicae* with wheat roots indicated that both attached and planktonic cells expressed the *nif* genes but with a 34–67-fold increase when in the root-associated state, indicating that permissive conditions had been achieved [69]. Since the oxygen-regulated *fix*NOP genes were upregulated at the roots, colonization of the root surface must have resulted in a microaerobic environment. These studies demonstrate that the tightly oxygen-regulated nitrogenase is expressed in and on plants, and therefore conditions are conducive for N-fixation.

2.2 Relative Importance of Hormone Production and Biological N-Fixation

Another common argument against endophytic N-fixation is that diazotrophic endophytes also produce phytohormones and have other plant growth-promoting properties that supersede effects of BNF. When plants are N-starved, they are stunted and chlorotic. The law of conservation of mass states that mass cannot be created, and since plant mass is at least 1.5% N, the observed increases in plant mass by inoculation with diazotrophic endophytes cannot be wholly explained by hormone production. A direct test of the involvement of BNF in the observed plant growth is through comparing the effects of inoculation with wild-type and nitrogenase mutant strains. Such experiments have been performed by several groups with mixed results. In the 1980s, *nif* mutants of *Azospirillum* still promoted plant growth [70]. However, no biological N-fixation (BNF) studies were done with the strains and it was not under sterile conditions. Only experiments in a gnotobiotic environment can rule out the effects of compensation by other diazotrophs. In such a study, a *nif*K mutant of *Azoarcus* strain BH72 was unable to promote growth of Kallar grass compared to inoculation with the wild-type strain endophyte, resulting in less biomass and less total N [13]. Expression of the nitrogenase gene was confirmed in the wild-type inoculated plants and was not in the mutant and the uninoculated control plants, further supporting that the source of the N was from the wild-type strain. In an earlier study using rice with the same *Azoarcus* strains, the plants had the same biomass and total protein content whether they were inoculated with wild-type or mutant strains [71]. However, the plants were not tested for BNF activity or confirmed to lack nitrogenase when inoculated with the mutant strain so it was unclear if reversion of the mutant phenotype or contamination with wild type had occurred. Addition to gnotobiotic sugarcane plantlets of wild type and a *nif* mutant of *Gluconacetobacter diazotrophicus* showed that the increased growth and total N content from inoculation of the wild-type strain were likely due to BNF [35]. In greenhouse studies with wheat, only inoculation with the wild-type strain of *K. pneumoniae* resulted in increased height and greenness under N-limited conditions compared to inoculation with a *nif* mutant or killed control strain, or uninoculated [36]. These studies indicate that, while phytohormone modulation, vitamin synthesis, and increased mineral uptake and stress tolerance conferred by diazotrophic endophytes are important [70], N-fixation is also a key factor in the benefit of inoculation.

2.3 Variability

While many diazotrophic endophyte strains have a broad host range in terms of plant species, there is variability in the results leading to disputes of the validity of endophytic N-fixation. The amount of N fixed and the biomass increase is dependent on the plant genotype and environmental condition [70]. Using the ^{15}N dilution technique or natural ^{15}N abundance, many studies have shown that 50–80% of the N in Brazilian sugarcane can be attributed to BNF [28, 72–75]. Studies by the International Rice Institute with rice showed that about 20–25% of N needs could be met through BNF [76]. Despite these significant levels for grasses, there are also many cases where there was no apparent contribution through BNF and no response to inoculation of diazotrophic strains [70]. For example, the commercially available mix of sugar cane endophytes, *Herbaspirillum seropedicae* and *Azospirillum brasilense*, enhanced the growth of only 3 of 30 genotypes of the model C4 grass, *Setaria viridis* [67]. In a similar study, 20 genotypes of the model C3 grass, *Brachypodium distachyon,* were tested for response to the inoculation but only four genotypes responded with increased growth under N-deficient conditions [77]. The ineffective interactions were not due to decreased root colonization; therefore the block is not at early recognition of the endophytes but rather at a later step in the symbiosis yet to be elucidated.

In addition to variability between plant varieties, there can also be variation of N-fixation within individual plants. Since endophytes are not limited to specific structures but may be found throughout the plant body, different cuttings of the same individual can have widely variable endophytic populations and N-fixation rates within them [78]. Cuttings of wild *Populus trichocarpa* (black cottonwood) showed zero to high levels of N-fixation as determined by ^{15}N incorporation and the acetylene reduction assay [78]. Diazotrophic endophyte research has been primarily on grasses so this phenomenon had not been previously reported. It remains to be seen if this variability is common to other eudicots due to their different structures (nodes, internodes, petioles, leaves, stems, and roots), and subsequent different niches, compared to monocots. Furthermore, since the poplar endophytes tended to be in microcolonies within the plant, it may be that specific populations and numbers must be attained for N-fixation to occur.

2.4 Insignificant Levels of N-Fixation

While the Rhizobium-legume symbiosis is thought to meet the N needs of the plants, it is viewed that associative N-fixation generally cannot [3, 57, 70]. However, legume crops are often still fertilized to maximize yields, any reduction in the need for chemical fertilization would be a substantial cost savings and have positive environmental impact, and, in natural ecosystems, plants can thrive without any additional inputs. These arguments are discussed below.

While there is little question that legumes in general have an evolutionarily superior symbiosis with N-fixing rhizobia, there have been losses in BNF capacity in modern cultivars [5, 79], leading to an increase in use of fertilizer. So if legume crops require fertilizer, then the levels of BNF from diazotrophic endophytes with non-legumes combined with limited fertilizer may be acceptable. In poor soils, even minor increases in N availability can improve crop yields [3]. Since the cost of fertilizer is linked to the cost of fossil fuels required for its production, the instability of pricing makes any reduction in the total need for N fertilizer a financial advantage. Likewise, considering the environmental costs of continued overuse of chemical fertilizers, the use of associative diazotrophs for crops would be beneficial even if the fertilizer use is not eliminated but only reduced [5].

A common argument against the significance of endophytic N-fixation is that there may not be active transfer of fixed N but rather only indirect transfer via death of the cells [80]. Such mineralization would be inefficient and delayed compared to active release of fixed N by living bacteria [81]. Also, the N in the mass of the dead bacterial cells is unlikely to be sufficient to explain the increased mass of inoculated plants. Mutants of associative diazotrophs can be isolated that have profound impacts on plant growth, fully rescuing the N-deficient phenotype [67]. Plants inoculated with an ammonium-secreting mutant of *Azospirillum* and grown under N limitation exhibited similar metabolic behavior as control plants grown under normal N conditions. This is perhaps the strongest argument that N fixed by the bacteria is incorporated by the plant and directly affects overall plant metabolism. While this is the case for a mutant strain, it is likely that under natural selection in N-limited environments, similar strains would exist or communities of strains would have such effects.

Where perennial crops have not been fertilized and in natural ecosystems, BNF is a significant source of N. In addition to the sugar cane examples described above, perennial elephant grass (*Pennisetum purpureum* Schum.) receives up to 70% of its N through BNF [82]. The N levels in coniferous forests cannot be explained by BNF only from nodulated plants that are restricted to open or riparian areas but must be from additional sources [83]. Endophytic N-fixation has indeed been quantified in *Pinus flexilis* [83] and *P. contorta* [84]. Using the ^{15}N dilution technique, *Paenibacillus polymyxa* was shown to provide up to 66% of the foliar N of lodgepole pine through BNF [84]. In boreal forests, epiphytic cyanobacteria of feather mosses provided up to 50% of the total N input [85, 86]. In primary succession sites where limited N sources are available, it is common to find plant species that do not have root nodule associations and yet thrive. *Populus* (poplar) and *Salix* (willow) are pioneer plant genera that dominate the rocky riparian zones of flood zones [87, 88]. Cuttings of wild *Populus trichocarpa* plants demonstrated N-fixation activity [78]. Through an understanding of the natural plant-microbe associations that allow high plant production in primary substrates, we may find solutions for dramatically reducing N fertilizer needs.

2.5 A Path Forward

If the dogma that only N-fixation in root nodules is significant for plant growth can be overcome, then research funding into the mechanisms of endophytic N-fixation will make possible the optimization of this important technology for more sustainable agriculture, forestry, and bioenergy. Overcoming dogma, however, can be a major hurdle. As described in the symposium, "Dispelling Dogmas," at the American Society for Microbiology symposium in 2014, ideas that *Mycobacterium tuberculosis* can be readily grown in culture, that viruses can be very large, and that bacteria have internal compartments were ridiculed [89]. It took scientists 30 years to accept that bacteria can communicate with each other, and now it is understood that quorum sensing is a common phenomenon [90]. If the data on endophytic N-fixation are viewed with a sense of scientific curiosity rather than contempt, the cases where diazotrophs did and did not promote plant growth could be compared and the mechanisms elucidated.

While N-fixation has been well studied in rhizobia and in specific free-living diazotrophs, the same depth of research has yet to be done for diazotrophic endophytes. Through transposon mutagenesis, either classically [91] or using high-throughput bioinformatics approaches such as TnSeq [92], all the genes required for N-fixation could be identified. Such studies could determine how endophytes resolve the "paradox" of the inhibition of nitrogenase with oxygen and the requirement by some strains for oxygen to generate the high levels of ATP required for N_2 fixation [2, 52]. Transcriptomics, RNA-seq, or metabolomics could be used to identify the form of N transferred from the endophyte to the plant. Understanding the regulation of N-fixation in endophytes will be important for comparisons to rhizobia. Control of N-fixation in response to N availability is less important in symbiotic bacteroids since rhizobia are committed to provide fixed N for the benefit of the plant [62]. If N-fixation is similarly regulated in endophytes in planta, it would suggest a symbiotic relationship.

Host specificity is a major question that must be addressed. While in some cases, cultivars seemed to be equally well colonized, only some responded to inoculations with increased growth under N-limitation. Since most studies are not performed under sterile conditions, it may be that the host microbiota are influencing the outcome. In addition, the genotype of the plant is also likely to be a major factor. While wild relatives would have co-evolved with the available microbiota, selecting for the most beneficial partnerships, modern cultivars were bred for optimum performance under artificial conditions [93]. In so doing, breeders may have lost the crop genes required for effective plant-microbe communication. Knowledge of the genes required for appropriate signaling and establishment of the symbiosis could aid in marker-assisted breeding of crop plants [12].

Although BNF can contribute substantial levels of N, endophyte-colonized crops still need N fertilizer [81] so there is need for more research to identify the best strains. Consortia of strains rather than individual strains could be optimized for improving crop yields, but this is a complex endeavor [94]. Microbe-microbe interactions can be beneficial or inhibitory depending on the strains. But using endophytes rather than rhizospheric bacteria as biofertilizers does have advantages since the plant endosphere is a more controlled environment with less competition [47]. In natural, productive ecosystems, microbial communities are co-dependent for effective nutrient cycling, so groups of strains from the same plant species in the same environment may increase the chance of success in synthetic communities. For example, a consortium of seven endophytic strains primarily from wild poplar to a hybrid poplar clone provided an estimated 65% of N from BNF, and increased leaf chlorophyll and root mass [95]. Choosing the most effective mix of strains may require molecular analysis to identify the key groups of strains from natural systems. One method is to analyze the core microbiome especially of plants grown under stress to identify the functionally important species, and then test the ability of the synthetic microbial community not only to colonize but also to persist [96]. By conducting a *nifH* gene expression analysis in sugar cane, it was shown that rhizobial species rather than the more abundant diazotrophic species may be the primary N-fixers in this species [97, 98]. Using mass spectrometry combined with advanced imaging techniques and sequencing may be necessary to identify the primary bacterial species that actually transfer fixed N to the host plant. For example, in a marine microbial mat system, isotope ratio mass spectroscopy (IRMS), nanoscale secondary ion MS (Nano-SIMS), catalyzed reporter deposition fluorescent in situ hybridization (CARD-FISH), and *nifH* clone library sequencing were all used to determine which cyanobacterial species was responsible for N-fixation within the microbial mat [99]. Using a suite of the latest advanced technologies, the most effective diazotrophic endophytes can be identified to help combat N-deficiency with less reliance on chemical fertilizers, leading to more environmentally and economically sustainable agriculture, forestry, and bioenergy crop production.

References

1. Downie JA. Legume nodulation. Curr Biol. 2014;24:R184–90.
2. Marchal K, Vanderleyden J. The "oxygen paradox" of dinitrogen-fixing bacteria. Biol Fertil Soils. 2000;30:363–73.
3. Stokstad E. The nitrogen fix. Science. 2016;353:1225–7.
4. Aneja VP, Blunden J, James K, Schlesinger WH, Knighton R, Gilliam W, Jennings G, Niyogi D, Cole S. Ammonia assessment from agriculture: U.S. status and needs. J Environ Qual. 2008;37:515–20.
5. Olivares J, Bedmar EJ, Sanjuan J. Biological nitrogen fixation in the context of global change. Mol Plant-Microbe Interact. 2013;26:486–94.
6. Venkateshwaran M, Volkening JD, Sussman MR, Ane JM. Symbiosis and the social network of higher plants. Curr Opin Plant Biol. 2013;16:118–27.
7. Froussart E, Bonneau J, Franche C, Bogusz D. Recent advances in actinorhizal symbiosis signaling. Plant Mol Biol. 2016;90:613–22.
8. Masson-Boivin C, Giraud E, Perret X, Batut J. Establishing nitrogen-fixing symbiosis with legumes: how many rhizobium recipes? Trends Microbiol. 2009;17:458–66.
9. Santi C, Bogusz D, Franche C. Biological nitrogen fixation in non-legume plants. Ann Bot. 2013;111:743–67.

10. Shulz B, Boyle C. What are endophytes? Soil Biol. 2005;9:1–13.
11. Sachs JL, Mueller UG, Wilcox TP, Bull JT. The evolution of cooperation. Q Rev Biol. 2004;79:135–60.
12. Carvalho TL, Ballesteros HG, Thiebaut F, Ferreira PC, Hemerly AS. Nice to meet you: genetic, epigenetic and metabolic controls of plant perception of beneficial associative and endophytic diazotrophic bacteria in non-leguminous plants. Plant Mol Biol. 2016;90:561–74.
13. Hurek T, Handley LL, Reinhold-Hurek B, Piche Y. Azoarcus grass endophytes contribute fixed nitrogen to the plant in an unculturable state. Mol Plant-Microbe Interact. 2002;15:233–42.
14. Welbaum GE, Sturz AV, Dong Z, Nowak J. Endophytic nitrogen fixation in sugarcane: present knowledge and future applications. Plant Soil. 2003;252:139–49.
15. Peng G, Zhang W, Luo H, Xie H, Lai W, Tan Z. Enterobacter oryzae sp. nov., a nitrogen-fixing bacterium isolated from the wild rice species Oryza latifolia. Int J Syst Evol Microbiol. 2009;59:1650–5.
16. Caballero-Mellado J, Martinez-Aguilar L, Paredes-Valdez G, Estrada-de Los Santos P. *Burkholderia unamae* sp. nov., an N_2-fixing rhizospheric and endophytic species. Int J Syst Evol Microbiol. 2004;54:1165–72.
17. Montanez A, Abreu C, Gill PR, Hardarson G, Sicardi M. Biological nitrogen fixation in maize (Zea mays L.) by N-15 isotope-dilution and identification of associated culturable diazotrophs. Biol Fertil Soils. 2009;45:253–63.
18. Reiter B, Burgmann H, Burg K, Sessitsch A. Endophytic nifH gene diversity in African sweet potato. Can J Microbiol. 2003;49:549–55.
19. Lopez BR, Bashan Y, Bacilio M. Endophytic bacteria of Mammillaria Fraileana, an endemic rock-colonizing cactus of the southern Sonoran Desert. Arch Microbiol. 2011;193:527–41.
20. Davis SC, Parton WJ, Dohleman FG, Smith CM, Del Grosso S, Kent AD, DeLucia EH. Comparitive biogeochemical cycles of bioenergy crops reveal nitrogen-fixation and low greenhouse gas emissions in a Miscanthus x giganteus agro-ecosystem. Ecosystems. 2010;13:144–56.
21. DeLuca TH, Zackrisson O, Gundale MJ, Nilsson MC. Ecosystem feedbacks and nitrogen fixation in boreal forests. Science. 2008;320:1181.
22. Dalton DA, Kramer S, Azios N, Fusaro S, Cahill E, Kennedy C. Endophytic nitrogen fixation in dune grasses (Ammophila arenaria and Elymus mollis) from Oregon. FEMS Microbiol Ecol. 2004;49:469–79.
23. Jimenez-Salgado T, Fuentes-Ramirez LE, Tapia-Hernandez A, Mascarua-Esparza MA, Martinez-Romero E, Caballero-Mellado J. *Coffea arabica* L., a new host plant for *Acetobacter diazotrophicus*, and isolation of other nitrogen-fixing acetobacteria. Appl Environ Microbiol. 1997;63:3676–83.
24. Rout ME, Chrzanowski TH, Westlie TK, DeLuca TH, Callaway RM, Holben WE. Bacterial endophytes enhance competition by invasive plants. Am J Bot. 2013;100:1726–37.
25. Doty SL, Dosher MR, Singleton GL, Moore AL, van Aken B, Stettler RF, Strand SE, Gordon MP. Identification of an endophytic *Rhizobium* in stems of *Populus*. Symbiosis. 2005;39:27–36.
26. Doty SL, Oakely B, Xin G, Kang JW, Singleton G, Khan Z, Vajzovic A, Staley JT. Diazotrophic endophytes of native black cottonwood and willow. Symbiosis. 2009;47:23–33.
27. Xin G, Zhang G, Kang JW, Staley JT, Doty SL. A diazotrophic, indole-3-acetic acid-producing endophyte from wild cottonwood. Biol Fertil Soils. 2009;45:669–74.
28. Boddey RM. Biological nitrogen fixation in sugarcane: a key to energetically viable biofuel production. CRC Crit Rev Plant Sci. 1995;14:263–79.
29. Reinhold-Hurek B, Hurek T. Life in grasses: diazotrophic endophytes. Trends Microbiol. 1998;6:139–44.
30. Boddey RM, Polidora JC, Resende AS, Alves BJR, Urquiaga S. Use of the 15N natural abundance technique for the quantification of the contribution of N2 fixation to sugarcane and other grasses. Aust J Plant Physiol. 2001;28:889–95.
31. Vessey JK. Plant growth promoting rhizobacteria as biofertilizers. Plant Soil. 2003;255:571–86.
32. Dobereiner J. Biological nitrogen fixation in the tropics: social and economic contributions. Soil Biol Biochem. 1997;29:771–4.

33. Urquiaga S, Xavier RP, de Morais RF, Batista RB, Schultz N, Leite JM, Sa JM, Barbosa KP, de Resende AS, Alves BJR, Boddey RM. Evidence from field nitrogen balance and 15N natural abundance data for the contribution of biological N2 fixation to Brazilian sugarcane varieties. Plant Soil. 2012;356:5–21.
34. de Morais RF, Quesada DM, Reis VM, Urquiaga S, Alves BJR, Boddey RM. Contribution of biological nitrogen fixation to Elephant grass (*Pennisetum purpureum* Schum.). Plant Soil. 2012;356:23–34.
35. Sevilla M, Burris RH, Gunapala N, Kennedy C. Comparison of benefit to sugarcane plant growth and $^{15}N_2$ incorporation following inoculation of sterile plants with *Acetobacter diazotrophicus* wild-type and *Nif-* mutant strains. Mol Plant-Microbe Interact. 2001;14:358–66.
36. Iniguez AL, Dong Y, Triplett EW. Nitrogen fixation in wheat provided by Klebsiella pneumoniae 342. Mol Plant-Microbe Interact. 2004;17:1078–85.
37. Elbeltagy A, Nishioka K, Sato T, Suzuki H, Ye B, Hamada T, Isawa T, Mitsui H, Minamisawa K. Endophytic colonization and in planta nitrogen fixation by a Herbaspirillum sp. isolated from wild rice species. Appl Environ Microbiol. 2001;67:5285–93.
38. Anand R. Endophytic colonization and nitrogen fixation by *Paenibacillus polymyxa* in association with lodgepole pine and western red cedar. Vancouver, BC: University of British Columbia; 2010.
39. Anand R, Chanway CP. N2-fixation and growth promotion in cedar colonized by an endophytic strain of Paenibacillus polymyxa. Biol Fertil Soils. 2013:235–9. doi:10.1007/s00374-012-0735-9.
40. Buren S, Young EM, Sweeny EA, Lopez-Torrejon G, Veldhuizen M, Voigt CA, Rubio LM. Formation of Nitrogenase NifDK tetramers in the mitochondria of Saccharomyces cerevisiae. ACS Synth Biol. 2017;6(6):1043–55.
41. Rogers C, Oldroyd GE. Synthetic biology approaches to engineering the nitrogen symbiosis in cereals. J Exp Bot. 2014;65:1939–46.
42. Khan Z, Guelich G, Phan H, Redman RS, Doty SL. Bacterial and yeast endophytes from poplar and willow promote growth in crop plants and grasses. ISRN Agron. 2012; doi:10.5402/2012/890280.
43. Knoth J, Kim S-H, Ettl G, Doty SL. Effects of cross host species inoculation of nitrogen-fixing endophytes on growth and leaf physiology of maize. GCB Bioenergy. 2013;5:408–18.
44. Kandel S, Herschberger N, Kim S-H, Doty SL. Diazotrophic endophytes of poplar and willow promote growth of rice plants in nutrient-limited conditions. Crop Sci. 2015;55:1765–72.
45. Khan Z, Kandel S, Ramos D, Ettl GJ, Kim S-H, Doty SL. Increased biomass of nursery-grown Douglas-fir seedlings upon inoculation with diazotrophic endophytic consortia. Forests. 2015;6:3582–93.
46. Mahanty T, Bhattacharjee S, Goswami M, Bhattacharyya P, Das B, Ghosh A, Tribedi P. Biofertilizers: a potential approach for sustainable agriculture development. Environ Sci Pollut Res Int. 2017;24:3315–35.
47. Timmusk S, Behers L, Muthoni J, Muraya A, Aronsson AC. Perspectives and challenges of microbial application for crop improvement. Front Plant Sci. 2017;8:49.
48. Gallon JR. The oxygen sensitivity of nitrogenase: a problem for biochemists and microorganisms. Trends Biochem Sci. 1981;6:19–23.
49. Mylona P, Pawlowski K, Bisseling T. Symbiotic nitrogen fixation. Plant Cell. 1995;7:869–85.
50. Boyd ES, Costas AM, Hamilton TL, Mus F, Peters JW. Evolution of molybdenum nitrogenase during the transition from anaerobic to aerobic metabolism. J Bacteriol. 2015;197:1690–9.
51. Poole RK, Hill S. Respiratory protection of nitrogenase activity in Azotobacter Vinelandii—roles of the terminal oxidases. Biosci Rep. 1997;17:303–17.
52. Gallon JR. Reconciling the incompatible - N-2 fixation and O-2. New Phytol. 1992;122:571–609.
53. Bergman B, Sandh G, Lin S, Larsson J, Carpenter EJ. Trichodesmium—a widespread marine cyanobacterium with unusual nitrogen fixation properties. FEMS Microbiol Rev. 2013;37:286–302.
54. Goebel NL, Turk KA, Achilles KM, Paerl R, Hewson I, Morrison AE, Montoya JP, Edwards CA, Zehr JP. Abundance and distribution of major groups of diazotrophic cyanobacteria and

their potential contribution to N(2) fixation in the tropical Atlantic Ocean. Environ Microbiol. 2010;12:3272–89.
55. Berman-Frank I, Lundgren P, Chen YB, Kupper H, Kolber Z, Bergman B, Falkowski P. Segregation of nitrogen fixation and oxygenic photosynthesis in the marine cyanobacterium Trichodesmium. Science. 2001;294:1534–7.
56. Setubal JC, dos Santos P, Goldman BS, Ertesvag H, Espin G, Rubio LM, Valla S, Almeida NF, Balasubramanian D, Cromes L, Curatti L, Du Z, Godsy E, Goodner B, Hellner-Burris K, Hernandez JA, Houmiel K, Imperial J, Kennedy C, Larson TJ, Latreille P, Ligon LS, Lu J, Maerk M, Miller NM, Norton S, O'Carroll IP, Paulsen I, Raulfs EC, Roemer R, Rosser J, Segura D, Slater S, Stricklin SL, Studholme DJ, Sun J, Viana CJ, Wallin E, Wang B, Wheeler C, Zhu H, Dean DR, Dixon R, Wood D. Genome sequence of Azotobacter vinelandii, an obligate aerobe specialized to support diverse anaerobic metabolic processes. J Bacteriol. 2009;191:4534–45.
57. James EK, Olivares FL. Infection and colonization of sugarcane and other graminacious plants by endophytic diazotrophs. Crit Rev Plant Sci. 1998;17:77–119.
58. Sprent JI, James EK. N2-fixation by endophytic bacteria: questions of entry and operation. In: Fendrik I, del Gallo M, Vanderleyden J, de Zamaroczy M, editors. Azospirillum VI and related microorganisms. Berlin: Springer; 1995. p. 15–30.
59. Flemming HC, Wingender J, Szewzyk U, Steinberg P, Rice SA, Kjelleberg S. Biofilms: an emergent form of bacterial life. Nat Rev Microbiol. 2016;14:563–75.
60. Dong Z, Zelmer CD, Canny MJ, McCully ME, Luit B, Pan B, Faustino RS, Pierce GN, Vessey JK. Evidence for protection of nitrogenase from O2 by colony structure in the aerobic diazotroph *Gluconacetobacter diazotrophicus*. Microbiology. 2002;148:2293–8.
61. Vessey JK, Pan B. Living a grounded life: growth and nitrogenase activity of Gluconacetobacter diazotrophicus on solid media in response to culture conditions. Symbiosis. 2003;35:181–97.
62. Dixon R, Kahn D. Genetic regulation of biological nitrogen fixation. Nat Rev Microbiol. 2004;2:621–31.
63. Agron PG, Ditta GS, Helinski DR. Oxygen regulation of nifA transcription in vitro. Proc Natl Acad Sci U S A. 1993;90:3506–10.
64. Hurek T, Egener T, Reinhold-Hurek B. Divergence in nitrogenases of Azoarcus spp., Proteobacteria of the beta subclass. J Bacteriol. 1997;179:4172–8.
65. Egener T, Hurek T, Reinhold-Hurek B. Endophytic expression of *nif* genes of *Azoarcus* sp. strain BH72 in rice roots. Mol Plant-Microbe Interact. 1999;12:813–9.
66. Chelius MK, Triplett EW. Immunolocalization of dinitrogenase reductase produced by Klebsiella pneumoniae in association with Zea mays L. Appl Environ Microbiol. 2000;66:783–7.
67. Pankievicz VC, do Amaral FP, Santos KF, Agtuca B, Xu Y, Schueller MJ, Arisi AC, Steffens MB, De Souza EM, Pedrosa FO, Stacey G, Ferrieri RA. Robust biological nitrogen fixation in a model grass-bacterial association. Plant J. 2015;81:907–19.
68. Monteiro RA, Balsanelli E, Wassem R, Marin AM, Brusamarello-Santos LCC, Schmidt MA, Tadra-Sfeir MZ, Pankievicz VC, Cruz LM, Chubatsu LS, Pedrosa FO, Souza EM. Herbaspirillum-plant interactions: microscopical, histological and molecular aspects. Plant Soil. 2012;356:175–96.
69. Pankievicz VC, Camilios-Neto D, Bonato P, Balsanelli E, Tadra-Sfeir MZ, Faoro H, Chubatsu LS, Donatti L, Wajnberg G, Passetti F, Monteiro RA, Pedrosa FO, Souza EM. RNA-seq transcriptional profiling of Herbaspirillum seropedicae colonizing wheat (Triticum aestivum) roots. Plant Mol Biol. 2016;90:589–603.
70. Dobbelaere S, Vanderleyden J, Okon Y. Plant growth-promoting effects of diazotrophs in the rhizosphere. Crit Rev Plant Sci. 2003;22:107–49.
71. Hurek T, Reinhold-Hurek B, Van MM, Kellenberger E. Root colonization and systemic spreading of Azoarcus sp. strain BH72 in grasses. J Bacteriol. 1994;176:1913–23.
72. Lima E, Boddey RM, Dobereiner J. Quantification of biological nitrogen fixation associated with sugar cane using ^{15}N aided nitrogen balance. Soil Biol Biochem. 1987;19:165–70.

73. Boddey RM, de Oliveira OC, Urquiaga S, Reis VM, de Olivares FL, Baldani VLD, Dobereiner J. Biological nitrogen fixation associated with sugar cane and rice: contributions and prospects for improvement. Plant Soil. 1995;174:195–209.
74. Boddey RM, Urquiaga S, Reis V, Dobereiner J. Biological nitrogen fixation associated with sugar cane. Plant Soil. 1991;137:111–7.
75. Urquiga S, Cruz KHS, Boddey RM. Contribution of nitrogen fixation to sugar cane: nitrogen-15 and nitrogen balance estimates. Soil Sci Soc Am J. 1992;56:105–14.
76. Roger PA, Ladha JK. Biological N2 fixation in wetland rice fields: estimation and contribuition to nitrogen balance. Plant Soil. 1992;141:41–55.
77. do Amaral FP, Pankievicz VC, Arisi AC, De Souza EM, Pedrosa F, Stacey G. Differential growth responses of Brachypodium distachyon genotypes to inoculation with plant growth promoting rhizobacteria. Plant Mol Biol. 2016;90:689–97.
78. Doty SL, Sher AW, Fleck ND, Khorasani M, Bumgarner R, Khan Z, Ko AWK, Kim SH, DeLuca TH. Variable nitrogen fixation in wild *Populus*. PLoS One. 2016;11:e0155979.
79. Graham, P. H., Hungria, M., and Tlusty, B. 2004. Breeding for better nitrogen fixation in grain legumes: Where do the rhizobia fit in? Online. Crop Management doi: 10.1094/CM-2004-0301-02-RV.
80. Bulgarelli D, Schlaeppi K, Spaepen S, Ver Loren van Themaat E, Schulze-Lefert P. Structure and functions of the bacterial microbiota of plants. Annu Rev Plant Biol. 2013;64:807–38.
81. Carvalho TL, Balsemao-Pires E, Saraiva RM, Ferreira PC, Hemerly AS. Nitrogen signalling in plant interactions with associative and endophytic diazotrophic bacteria. J Exp Bot. 2014;65:5631–42.
82. de Morais RF, Quesada DM, Reis VM, Urquiaga S, Alves BJR, Boddey RM. Contribution of biological nitrogen fixation to elephant grass (Pennisetum purpureum Schum.). Plant Soil. 2017;356:23–34.
83. Moyes AB, Kueppers LM, Pett-Ridge J, Carper DL, Vandehey N, O'Neil J, Frank AC. Evidence for foliar endophytic nitrogen fixation in a widely distributed subalpine conifer. New Phytol. 2016;210(2):657–68.
84. Bal A, Chanway CP. Evidence of nitrogen fixation in lodgepole pine inoculated with diazotrophic Paenibacillus polymyxa. Botany. 2012;90:891–6.
85. DeLuca TH, Zackrisson O, Nilsson MC, Sellstedt A. Quantifying nitrogen-fixation in feather moss carpets of boreal forests. Nature. 2002;419:917–20.
86. Rousk K, Jones DL, DeLuca TH. Moss-cyanobacteria associations as biogenic sources of nitrogen in boreal forest ecosystems. Front Microbiol. 2013;4:150.
87. Isebrands JG, Richardson J. Poplars and willows: trees for society and the environment. Rome: FAO/CAB; 2014.
88. Stettler RF. Cottonwood and the river of time: on trees, evolution, and society. Seattle: University of Washington Press; 2009.
89. Fox JL. Skepticism, curiosity helped to overcome microbiological dogmas. Microbe. 2014;9:355–6.
90. Fuqua C, Greenberg EP. Listening in on bacteria: acyl-homoserine lactone signalling. Nat Rev Mol Cell Biol. 2002;3:685–95.
91. Jacobs MA, Alwood A, Thaipisuttikul I, Spencer D, Haugen E, Ernst S, Will O, Kaul R, Raymond C, Levy R, Chun-Rong L, Guenthner D, Bovee D, Olson MV, Manoil C. Comprehensive transposon mutant library of Pseudomonas aeruginosa. Proc Natl Acad Sci U S A. 2003;100:14339–44.
92. Wetmore KM, Price MN, Waters RJ, Lamson JS, He J, Hoover CA, Blow MJ, Bristow J, Butland G, Arkin AP, Deutschbauer A. Rapid quantification of mutant fitness in diverse bacteria by sequencing randomly bar-coded transposons. MBio. 2015;6:e00306–15.
93. Perez-Jaramillo JE, Mendes R, Raaijmakers JM. Impact of plant domestication on rhizosphere microbiome assembly and functions. Plant Mol Biol. 2016;90:635–44.
94. Rosier A, Bishnoi U, Lakshmanan V, Sherrier DJ, Bais HP. A perspective on inter-kingdom signaling in plant-beneficial microbe interactions. Plant Mol Biol. 2016;90:537–48.

95. Knoth JL, Kim SH, Ettl GJ, Doty SL. Biological nitrogen fixation and biomass accumulation within poplar clones as a result of inoculations with diazotrophic endophyte consortia. New Phytol. 2014;201:599–609.
96. Busby PE, Soman C, Wagner MR, Friesen ML, Kremer J, Bennett A, Morsy M, Eisen JA, Leach JE, Dangl JL. Research priorities for harnessing plant microbiomes in sustainable agriculture. PLoS Biol. 2017;15:e2001793.
97. Burbano CS, Liu Y, Rosner KL, Reis VM, Caballero-Mellado J, Reinhold-Hurek B, Hurek T. Predominant *nifH* transcript phylotypes related to *Rhizobium rosettiformans* in field-grown sugarcane plants and in Norway spruce. Environ Microbiol Rep. 2011;3:383–9.
98. Thaweenut N, Hachisuka Y, Ando S, Yanagisawa S, Yoneyama T. Two seasons' study on *nifH* gene expression and nitrogen fixation by diazotrophic endophytes in sugarcane (*Saccharum* spp. hybrids): expression of *nifH* genes similar to those of rhizobia. Plant Soil. 2011;338:435–49.
99. Woebken D, Burow LC, Prufert-Bebout L, Bebout BM, Hoehler TM, Pett-Ridge J, Spormann AM, Weber PK, Singer SW. Identification of a novel cyanobacterial group as active diazotrophs in a coastal microbial mat using NanoSIMS analysis. ISME J. 2012;6:1427–39.

Chapter 3
Endophyte-Promoted Nutrient Acquisition: Phosphorus and Iron

Sagar Chhabra and David N. Dowling

3.1 Introduction

Phosphorus (P) and iron (Fe) are essential nutrients required by plants; however the bio-availability of both these macro- and micronutrients is low in soil as both P and Fe form insoluble mineral complexes; for example, iron is generally present as a Fe^{3+} (ferric ion) complex with oxyhydroxide polymers in soil and is not bio-available under alkaline pH conditions [1], whereas phosphorus complexes with calcium, iron, or aluminum in soil under alkaline or acidic pH conditions and these are not directly available in the form of orthophosphate anions for plant uptake [2, 3]. Plants have adapted to low mineral nutrient environments by using several strategies to overcome nutrient deficiency and increase plant nutrient uptake. These include inducing morphological or physiological changes to the root-soil interface by changing plant root architecture such as extensive root branching [4–6]. Increase in length of root and root hairs and root angle can also increase the spatial access and availability of nutrients present in soil to plants [7–10]. The increase in physiological or biochemical activities such as phytosiderophore production, organic anion production, and excretion of protons and increase in hydrolytic enzymes, e.g., phosphatase or phytase activity, are all associated with an increase in nutrient acquisition of either Fe or P by plants [4, 6, 11].

The improved availability of plant nutrients has long been associated with plant microbial interactions, in particular, arbuscular mycorrhizal fungi (AMF) associations that are involved in the transport or acquisition of P and also Fe in plants [12, 13]. The presence of microorganisms other than AMF associated with plants such as

S. Chhabra
MicroGen Biotech Ltd, ERIC Centre, Institute of Technology Carlow, Carlow, Ireland

D.N. Dowling (✉)
EnviroCore, Dargan Centre, Institute of Technology Carlow, Carlow, Ireland
e-mail: dowlingd@itcarlow.ie

S.L. Doty (ed.), *Functional Importance of the Plant Microbiome*,
DOI 10.1007/978-3-319-65897-1_3

bacteria and fungi present in rhizosphere soil or bacterial and fungal endophytes that occur asymptomatically in plant organs and tissues has also been shown to provide benefits to plant health by nutrient acquisition [14–17]. In this chapter we focus on the importance of endophyte microorganisms with respect to their role in P and Fe nutrient acquisition in plants.

3.2 Microorganism Functions Implicated in Nutrient Fe and P Uptake in Plants

Plant-associated microorganisms (the plant microbiome) are commonly attributed with a range of plant growth promotion functions such as biological nitrogen fixation, phosphate solubilization, production of siderophores, ACC deaminase activity, production of phytohormones, and biocontrol activity [18, 19]. The plant and its associated microbiome have been termed the holobiont [20] and the plant microbiome is influenced by soil type and plant genome [21, 22]. Microorganisms can help increase nutrient Fe or P uptake and benefit plants directly due to microbial siderophore production or phosphate mineralization or solubilization activity. Other plant growth promotion traits such as plant hormone production or increasing plant stress tolerance by reducing plant ethylene levels by ACC deaminase activity or biocontrol functions may also help increase plant growth by increasing the soil root interface, thus indirectly increasing Fe and P nutrient uptake in the plant [23–25]; see Fig. 3.1.

The early interaction of microorganisms with land plants in the form of mycorrhizal fungal associations (AMF) is hypothesized to have evolved from fungal endophytes that developed external hyphae to provide plant nutrient support to plants in exchange for enriched carbon sources available from the host plant [26]. The AMF association with plants is the oldest and most widely represented on land [26, 27]. AMFs function by scavenging of P and Zn nutrients from soil but are also known to enhance acquisition of nutrients such as Fe, Ca, K, and S in plants [12]. Besides the AMF interactions, the other widely recognized group of fungi associated with plants are the non-clavicipitaceous group of Class 4 endophytes also known as dark septate endophytes (DSEs) [28]. The DSEs are known to be present in over 600 different plant species and are found worldwide [29]. The DSEs can help improve phosphorous supply in plants and in certain conditions appear to replace AMFs and ectomycorrhizal fungi at sites with extreme environmental conditions [28]. Among the other fungi, the basidiomycete fungus *Piriformospora indica*, a recently recognized endophyte, was shown to be distributed over a broad geographical area and interact with a number of angiosperms (around 145 or more) including the model plant *Arabidopsis thaliana* and with certain other members of the *Brassicaceae* family where AMF infections or associations are not detected. *P. indica* stimulates nutrient uptake in the roots [30, 31] and solubilizes insoluble phosphate in plants [32].

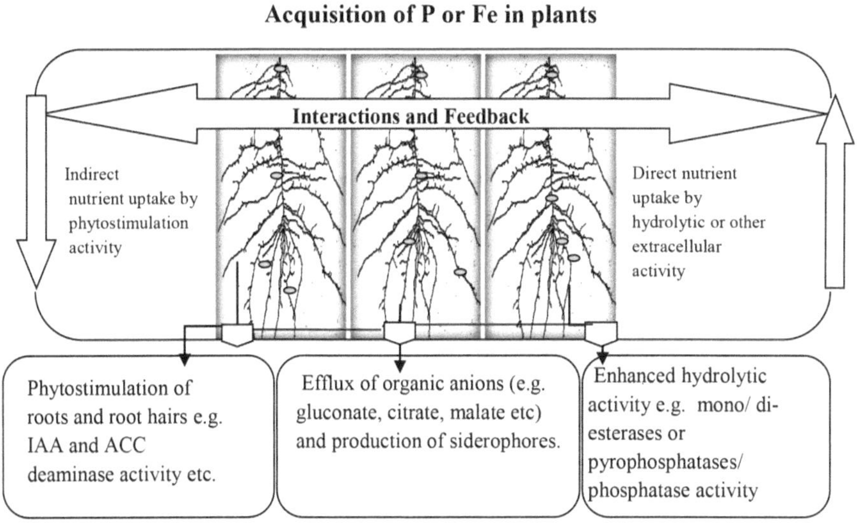

Fig. 3.1 Microbial functional aspects in plants that impact nutrient P or Fe availability

The presence of large numbers of endophyte bacteria isolated from the plants' microbiota, for example, *Gluconacetobacter*, *Azospirillum*, *Azotobacter*, *Bacillus*, *Burkholderia*, *Herbaspirillum*, *Pseudomonas*, *Achromobacter*, *Klebsiella*, *Chryseobacterium*, and *Pantoea* genera, has been observed to improve plant growth through stimulation of root development [33–35]. The microbial isolates such as *Pantoea*, *Pseudomonas*, *Citrobacter*, *Azotobacter*, *Streptomyces*, or other newly recognized groups of bacteria have also been identified as contributing to plant growth promotion by virtue of nutrient acquisition traits [33, 34, 36].

3.3 P and Fe Transport in Plants and Plant-Associated Microorganisms

The plant transports both Fe and P in response to nutrient deficiency conditions and there are several P or Fe transporters characterized both in plants and microorganisms [37–40]. Plant roots are the primary site for plant nutrient acquisition and under P- and Fe-depleted conditions undergo morphological changes in order to adapt to the changing nutrient condition or availability in soil [4, 11]. An increase in acidification of the rhizosphere environment such as by exudation of proton or carboxylate ions such as citrate, malate, or oxalate can greatly enhance mobilization of P in plants such as by chelation or ligand exchange of P bound or complexed to Ca, Fe, or Al present in soil [13]. Secretion of phosphatases or phytases can mobilize organic P through hydrolysis and has been shown to increase P availability in plants [6, 13].

The presence of microorganisms associated with plants is known to increase P availability in plants. Microorganisms such as bacteria and fungal endophytes isolated from plants have been shown under in vitro conditions to be involved in mineral phosphate solubilization activity by acidification of the extracellular environment, and by production of organic acid anions such as gluconic acid [41], malic acid [42], citric acid [43, 44], salicylic acid, and benzeneacetic acid [43]. Microorganisms are also hypothesized to be involved in other relevant activities such as proton extrusion and by ammonium ion assimilation that are linked with mineral phosphate solubilization activity [45]. The organic phosphate mineralization by microorganisms involves phosphatase activity, e.g., acid or alkaline phosphatase activity or phytase activity, which may contribute to availability of inorganic phosphate for plant uptake [46, 47].

The transport of P in both plants and microorganisms is mainly associated with two transporters, the high-affinity Pi transport systems and low-affinity Pi transport systems [13, 38]. The high-affinity Pi transporter in plants is the major transporter family responsible for transport of P in roots or in cells with close contact to the soil matrix. The low-affinity Pi transporters are mainly active in vascular tissues and involved in the internal distribution and re-mobilization of P [48]. Phosphate transport in plants by the high-affinity transporter system is H^+/ATP dependent and is activated or expressed when the external P level in plants or cells in close contact to the soil matrix is low. The high-affinity phosphate transporter system is grouped within the major Pht1 family and has shared topology among fungi, yeast, plant, and animal Pht1 transporters [49, 50]. The high-affinity phosphate transporters have been characterized in a number of plant and fungal species; however, the role of phosphate transporters among certain AMFs has not been verified due to the lack of a stable transformation system [51]. A study on the P transporter of the fungal endophyte *Piriformospora indica* [51] identified a high-affinity phosphate transporter PiPT belonging to the major facilitator superfamily (MFS) found in bacteria which is also conserved in eukaryotes [52]. The study also recognized the structural/functional relationships of Pi/H^+ symporters and the proton motive force driving the translocation of Pi in the host plant by the basidiomycete fungus under the Pi limitation condition [51, 52].

Iron transport or acquisition in plants involves two strategies and is dependent on the plant type under iron-deficient conditions. The acidification of the extracellular soil environment by proton extrusion and reduction of chelated Fe^{3+} by ferric chelate reductase at the plant root surface enhance bio-availability of Fe as ferrous (Fe^{2+}) ion in nongraminaceous monocotyledonous and in dicotyledonous plants. However, production of mugineic acid dependent phytosiderophores is an important mechanism for Fe chelation and availability as ferrous iron (Fe^{2+}) for transport or acquisition of Fe in monocotyledons, especially among grasses [53]. The uptake of Fe by microorganisms involves a similar strategy to that of plants and involves chelation of unavailable Fe by specific or a range of siderophores and/or the use of reductases which help to increase available extracellular Fe for uptake by the microorganism [54]. Iron uptake in both plants and microorganisms involves Fe transporters and there is considerable similarity in certain cases between some microorganism and plant Fe transporters such as in the case of yeast and specific plant Fe transporters [55, 56]. In conclusion there are a number of key plant growth promotion functional traits that are associated with microorganisms involved in P and Fe availability and these are summarized in Table 3.1.

3.4 Functional Role of P and Fe Bacterial and Fungal Endophytes

3.4.1 P Endophytes and their Role (Fungal and Bacterial)

The P endophytes or other associated P microorganisms increase Pi availability in soils by mineral phosphate solubilization or by organic phosphate mineralization activity. The mechanistic basis or direct involvement of the P endophyte to increase plant growth or biomass has been demonstrated in studies under P limitation that is discussed below and summarized in Table 3.2.

Mineral phosphate solubilization is an effective strategy for the provision of P to the plant. In a study by Crespo et al. [70] they identified that the ability to solubilize inorganic phosphate was associated with acidification of the plant root environment by plant root-associated bacteria. The root of wheat and tomato was colonized by the bacterial endophyte *Gluconacetobacter diazotrophicus* previously isolated from sugarcane [82] that efficiently enhanced acidification of the plant root by production of gluconic acid. In contrast, the gluconic acid biosynthesis gene mutant (PQQ-GDH) of *G. diazotrophicus* lacked the ability to acidify in a test medium with plants, thus underlying the functional role of this trait.

The direct role of P endophytes in plant growth promotion have also been described in other studies, for example, Kumar et al. [74] in a study on maize colonized by the fungal endophyte *P. indica* reported a higher increase in biomass of plants under P limiting conditions. The difference in biomass between colonized and non-colonized plants was a 2.5-fold increase at limiting P and 1.2-fold increase at non-limiting P conditions respectively, thus underlying the function of the endophyte to be more effective under the P deficient condition. Li et al. [75, 76] in a study on perennial grass *Achnatherum sibiricum* infected by the fungal species *Neotyphodium* sp. recognized a significant increase in acid phosphatase activity under P deficient and N non-limiting conditions. The biomass of the endophyte infected plant was not affected by P limitation and was similar to plants grown under non-limiting P or N conditions. Malinowski et al. [80] found that an infected *Festuca arundinacea* (tall fescue) with the endophyte *Neotyphodium coenophialum* under P limiting conditions expressed an increased root absorption area through reduced root diameter and increased root hair length compared with the endophyte free counterpart. Altered root diameter and root hair length in this study was associated with the functional role by the endophyte present in tall fescue.

The role of fungal P endophytes to increase plant growth and to enhance phosphorous efficiency was also demonstrated by studies involving dark septate fungi (DSEs) present in the plant. Barrow and Osuna [79] reported an increase in shoot and root

Table 3.1 Examples of plant-associated endophyte microorganisms implicated in nutrient acquisition and other plant functions

Isolation source plant	Endophyte microorganism	Functional traits	Reference
Arachis hypogaea Peanut	*Pantoea agglomerans*	Mineral phosphate solubilization, siderophore production	[57]
Calophyllum brasiliense Guanandi	*Trichoderma* sp.	Mineral phosphate solubilization	[58]
Glycine max Soybean	*Enterobacter sakazakii*, *Pseudomonas straminae*, *Acinetobacter calcoaceticus*, *Pseudomonas* sp.	Mineral phosphate solubilization, IAA, biological nitrogen fixation	[59]
Glycine max Soybean	*Rhizoctonia* sp. *Fusarium verticillioides*	Phytase	[60]
Lippia sidoides Pepper-rosmarin	*Lactococcus lactis*	Calcium phosphate, phosphate mineralization activity-calcium phytate, solubilize/mineralize phosphate from poultry litter	[61]
Mammillaria fraileana Wild cactus	*Pseudomonas putida* M5TSA, *Enterobacter sakazakii* M2PFe, *Bacillus megaterium* M1PCa	Mineral phosphate solubilization	[62]
Manihot esculenta Cassava	*Pantoea dispersa*	Mineral phosphate solubilization, biological nitrogen fixation	[43]
Miscanthus giganteus Miscanthus	*Pseudomonas fluorescens*	Mineral phosphate solubilization	[41]
Moringa peregrine Moringa	*Sphingomonas* sp. LK18, *Methylobacterium radiotolerans* LK17, *Bacillus subtilis* LK14, *Bacillus subtilis* LK15, *Sphingomonas* sp. LK16	Mineral phosphate solubilization, acid phosphatase, IAA	[63]
Oryza sativa var. *Japonica* c.v. Rice	*Paenibacillus kribbensi*, *Bacillus aryabhattai*, *Klebsiella pneumoniae*, *Bacillus subtilis*, *Microbacterium trichotecenolyticum*	Biological nitrogen fixation, mineral phosphate solubilization, IAA	[64]
Pachycereus pringlei Giant cardon cactus	*Bacillus pumilus* var.2, *B. subtilis* var.2, *Actinomadura oligospora*, *Citrobacter* sp.	Mineral phosphate solubilization	[65]

(continued)

Table 3.1 (continued)

Isolation source plant	Endophyte microorganism	Functional traits	Reference
Panax ginseng Ginseng	*Lysinibacillus fusiformis*, *Bacillus cereus*, *B. megaterium*, *Micrococcus luteus*	Mineral phosphate solubilization IAA, siderophore production	[117]
Phaseolus vulgaris Common Bean/French bean	*Rhizobium endophyticum* sp. Nov	Phytate	[66]
Phaseolus vulgarus Common bean/French bean	*Pseudomonas* sp.	Mineral phosphate solubilization	[67]
Piper nigrum Black pepper	*Klebsiella* sp., *Enterobacter* sp.	Mineral phosphate solubilization siderophore production, ACC deaminase,IAA production	[68]
Pseudotsuga menziesii Douglas-fir	*Rhodotorula graminis*, *Acinetobacter calcoaceticus*, *Rhizobium tropici bv populus*, *Sphingomonas yanoikuyae*, *Pseudomonas putida*, *Rahnella* sp., *Burkholderia* sp., *Sphingomonas* sp.	Mineral phosphate solubilization, Siderophores production, biological nitrogen fixation	[69]
Saccharum officinarun Sugarcane	*Gluconacetobacter diazotrophicus*	Biological nitrogen fixation, mineral phosphate solubilization	[70]
Shorea leprosula and *Shorea selanica* Meranti	*Trichoderma spirale*	Mineral phosphate solubilization and inhibition of fungal pathogen (*fusarium*)	[71]
Triticum aestivum Wheat	*Streptomyces tricolor mhce0811*	Mineral phosphate solubilization phytase, siderophores, IAA, chitinase	[42]

biomass, and phosphorus use efficiency in *Atriplex canescens* by the fungal endophyte *Aspergillus ustus*. Jumpponen et al. [77] reported increased foliar P concentration and an increase in plant biomass of more than 50% following fungal inoculation and N amendment in endophyte-infected *Pinus contorta* by the fungal endophyte *Phialocephala fortini*. Newsham [78] recognized increased root, shoot, total P content, and total biomass and an increase in the number of tillers in endophyte-infected *Vulpia ciliata* by the fungal endophyte *Phialophora graminicola*.

Studies defining the mechanistic basis of P transport by endophytic microorganisms present in plants under P-deficient conditions also demonstrate the essential role of the endophyte. Hiruma et al. [73] in a study on an ascomycete fungal endophyte *Colletotrichum tofieldiae* (*Ct*) in *Arabidopsis* identified the role of this endophyte in

Table 3.2 Mechanisms of microbial phosphorous solubilization/transfer in endophyte-plant interactions

Endophyte	Original plant host	characteristics Or functional trait	Plant host/study	Effects	Reference	Comments
Gluconacetobacter diazotrophicus	*Saccharum officinarum* (sugarcane)	Transposon mutants in *pqq BC* and *E* and also *gdhA*	Not applicable In vitro study	A transposon mutant library identified gluconic acid synthesis as a required pathway for mineral phosphate (P) and zinc solubilization	[72]	However no direct link with the plant's regulatory network and P uptake in plants has been recognized to date
Gluconacetobacter diazotrophicus	*Saccharum officinarum* (sugarcane)	PQQ-GDH mutant MF105strain of *G. diazotrophicus*	*Lycopersicon esculentum* (tomato) *Triticum aestivum* (wheat)	Seedlings inoculated with a PQQ-GDH mutant MF105strain of *G. diazotrophicus* showed no acidification compared to WT	[70]	*G. diazotrophicus* produced solubilization only when aldoses were used as the C-source
Colletotrichum tofieldia Ct	*Arabidopsis thaliana*	^{32}P translocation experiments and transcriptional analysis of P transporter genes in *Arabidopsis*	*Arabidopsis thaliana*	*C. tofieldiae* (Ct) root Fungal endophyte transfers the macronutrient phosphorus to shoots, promotes plant growth, and increases fertility under phosphorus-deficient conditions The host's phosphate starvation response (PSR) system controls Ct root colonization and is needed for plant growth promotion (PGP)	[73]	Nine transporter genes were significantly upregulated in phosphate-starved roots in the absence of Ct while two, Pht1;2 and Pht1;3, were induced at higher levels only in the presence of Ct under Pi-limiting conditions, similar to mycorrhizal associations

Piriformospora indica	*Zea mays* (maize)	Knockdown gene encoding a phosphate transporter (PiPt) in *P. indica*	*Zea mays* (maize)	Higher amounts of phosphate were found in plants colonized with WT *P. indica* than that of non-colonized plants or plants with knockdown gene encoding a phosphate transporter (PiPt) in *P. indica*	[51]	Observation suggests that PiPt is actively involved in *P. indica* and helps improve the nutritional status of the status of the host plant
Piriformospora indica	*Zea mays* (maize)	^{32}P translocation experiments	*Zea mays* (maize)	Increase in biomass between colonized and non-colonized plants was 2.5 times that at limiting P. *P. indica* enhances growth more effectively at low P. Considerable amount of ^{32}P was measured in maize plants, suggesting that hyphae of *P. indica* were transporting P to host plant under P-limiting conditions	[74]	Speculated that phosphocholine effluxed by the fungus to the plant and pi would then be taken up by the plant via an H+ co-transporter. A phosphate transporter PiPT gene was not expressed in P-rich conditions, only under P-deprived condition

(continued)

Table 3.2 (continued)

Endophyte	Original plant host	characteristics Or functional trait	Plant host/study	Effects	Reference	Comments
Neotyphodium sp.	*Achnatherum sibiricum*		*Achnatherum sibiricum*	Under P deficiency and N availability, endophyte infection significantly increased acid phosphatase activity of endophyte-infected plants and the biomass of endophyte-infected plants was not affected under P deficiency, with similar growth as under available P or available N conditions	[75, 76]	Under both N- and P-deficient conditions, plant had greater P concentrations in roots and no significant difference in biomass was found between endophyte-infected and endophyte-free plants
Phialocephala fortini	*Pinus contorta*	Inoculated vs. non-inoculated plants	*Pinus contorta*	Increased foliar P concentration. A combination of N amendment and fungal inoculation increased host biomass by more than 50% beyond that obtained via N amendment alone	[77]	
Phialophora graminicola	*Vulpia ciliata* ssp. *ambigua*	Inoculated vs non-inoculated plants	*Vulpia ciliata* ssp. *ambigua*	Increased tillers, increased root, shoot, and total biomass. Increased root N content plus increased root, shoot, and total P, reduced shoot N	[78]	

Aspergillus ustus	*Atriplex canescens*	Experimental system to allow access to P only through the endophytic fungus	*Atriplex canescens*	*A. ustus* obtained plant carbon, increased shoot and root biomass, and phosphorus use efficiency	[79]	
Neotyphodium coenophialum	*Festuca arundinacea* (tall fescue)	*Neotyphodium* infected vs. non-infected plant	*Festuca arundinacea* (tall fescue)	*N. coenophialum* under P-limiting conditions expressed an increased root absorption area through reduced root diameter and increased root hair length compared with the endophyte-free counterpart	[80]	
Dark septate fungal sp. *(unknown)*	*Carex firma* and *Carex sempervirens*		*Carex firma* and *Carex sempervirens*	Significant increase in shoot P content in both the host plants. Increase in dry matter production detected in *Carex firma*	[81]	

transfer of phosphorus to *Arabidopsis* shoots. This study showed that the host's phosphate starvation response (PSR) system controls *Ct* root colonization and is needed for plant growth promotion, and also the role of *Ct*-mediated plant growth promotion was recognized to be mediated by the plant innate immune system. This study hypothesized that the *Ct* association in the host root of *A. thaliana* and other *Brassicaceae* members has essential components important for developing these associations that are usually absent in mycorrhizal symbiosis. The importance of the P endophyte and P transport function in plants has also been demonstrated by *P. indica* in maize plants. Yadav et al. [51] reported that higher amounts of phosphate were found in plants colonized with wild-type *P. indica* than that of non-colonized plants or plants with a knockdown phosphate transporter (PiPT). It was suggested that PiPT of *P. indica* was actively involved in phosphate transfer *in planta* and can improve the nutritional status of the host plant.

3.4.2 Fe Endophytes and their Role (Fungal and Bacterial)

The role of endophyte and other microorganisms in iron acquisition by plants is associated with siderophore production. There are over 500 different types of siderophores produced by microorganisms [37] and purified bacterial siderophore has been recently shown to restore growth to iron-limited and stunted tomato plants [83]. Siderophores produced by microorganisms not only directly improve Fe availability to microorganisms and plants by direct chelation from soil but can also increase iron availability based on their competition for Fe with other microorganisms and pathogens. Studies defining the Fe availability and plant growth or biocontrol function by Fe endophytes are summarized in Table 3.3.

The importance of the siderophore-producing trait by Fe endophytes is demonstrated by its direct role in increasing plant growth or by improvement of host fitness. Rungin et al. [89] in a study using a bacterial endophyte *Streptomyces* sp. previously isolated from jasmine rice (*Oryza sativa* L. cv. KDML105) and its siderophore mutant (*desD*) showed an enhancement of plant growth with a significant increase in plant biomass in rice (*Oryza sativa*) and mungbean (*Vigna radiata*) by the siderophore-producing endophyte. The increase in plant growth or biomass was higher in *Streptomyces*-treated plants producing siderophore compared to a siderophore-deficient *desD* mutant and untreated control plants, thus underlying the functional importance of siderophore in enhancement of growth in plants. Rosconi et al. [86] in a study on a serobactin-producing bacterial endophyte *Herbaspirillum seropedicae* responsible for Fe acquisition by the microorganism and with its uptake mutant (Hsero_2345 gene) in an experiment on rice (*Oryza sativa*) showed that serobactin-mediated iron acquisition contributes to competitive fitness in the host plant.

The role of siderophore produced by endophytic bacteria has also been demonstrated through its biocontrol function or synergistic role in plant growth promotion and colonization in certain studies. Verma et al. [88] in a study on bacterial

Table 3.3 Mechanisms of microbial iron solubilization/transfer in endophyte-plant interactions

Endophyte	Original plant host	Characteristics	Plant host	Effects	Reference	Comments
Epichloë festucae	*Festuca trachyphylla (Festuca longifolia)*	Epichloenin A (ferrichrome siderophore) *sidN* mutant defective in synthesis	*Lolium perenne* (ryegrass)	Fungal siderophore was detected by HRMS and NMR in plant guttation fluid (xylem sap) inoculated with fungus but not in mutant inoculated plants	[84]; [85]	Indicates fungal siderophore is produced in planta. Role in maintaining mutualistic nature of symbiosis
Herbaspirillum seropedicae Z67	*Oryza sativa* (rice)	Serobactin uptake mutant Hsero_2345 gene	*Oryza sativa* (rice)	Contributes to competitive fitness of endophyte within plant. WT is more competitive after 8 days	[86]	
Pseudomonas fluorescens PICF7	Olea europaea (olive)	Pyoverdine (pseudobactin) mutant (*pvdI*). Siderophore negative	*Olea europaea* (olive)	Does not affect external or internal root colonization Does not affect biological control *Verticillium* wilt of olive (VWO) (*Verticillium dahliae*)	[87]	Uptake of pseudobactin would not be affected. Did not test for competition with WT

(continued)

Table 3.3 (continued)

Endophyte	Original plant host	Characteristics	Plant host	Effects	Reference	Comments
Streptomyces azR-051	*Azadirachta indica*	Wt high-expressing siderophore and IAA strains	*Lycopersicon esculentum* (tomato)	PGPR and enhanced biocontrol of early blight disease caused by *Alternaria alternata* compared to stains with lower levels of IAA and siderophore	[88]	Not clear if IAA or siderophore is the key factor
Streptomyces sp. GMKU3100	*Oryza Sativa* (Thai jasmine rice)	High siderophore producer but no P sol or IAA production. *desD* mutant-siderophore negative	*Vigna radiata* (mung bean) *Oryza sativa* (rice)	Increase in plant biomass parameters with WT inoculation compared to siderophore mutant or uninoculated controls	[89]	
Chryseobacterium C138	*Oryza sativa* (Rice)	Siderophore extracted from this strain	*Lycopersicon esculentum* var. *Marglobe (tomato)*	Iron-starved tomato plants when treated with siderophore C138 were restored with respect to growth parameters to full-iron Hoagland solution	[83]	Purified bacterial siderophore C138 could supply Fe to iron-starved plants

Streptomyces endophytes recovered from *Azadirachta indica* reported significant plant growth promotion in tomato and biocontrol against *Alternaria alternata*, causal agent of early blight disease in tomato plants. A significant antagonistic activity by *Streptomyces* endophytes against the pathogen was linked to the high Fe complexing capacity of isolates (i.e., siderophore production). All isolates of *Streptomyces* prolifically produced IAA and siderophores demonstrating that both IAA and siderophores play a vital role in promotion of plant growth and in suppression of the pathogen. Koulman et al. [84] in a study using the fungal endophyte *Epichloë festucae* isolated from *Festuca trachyphylla* and its siderophore *sidN* mutant in an experiment on *Lolium perenne* reported the role of siderophore in colonization in xylem sap of *L. perenne*. Further work by Johnson et al. [85] demonstrated that this gene (*sidN*) played a key role in maintaining the mutualistic interaction with its host plant, highlighting the importance of iron homeostasis for the symbiotic interaction.

3.5 Perspectives on the Role of Endophytes in Fe and P Nutrient Acquisition: Potential Application for Agriculture and Future Prospects

The unavailability of both Fe and P in many soils had been recognized as a major growth-limiting factor in many agricultural systems [36, 90]. The inoculation of crops with specific microorganisms has the potential to reduce application rates of phosphate and can also improve iron uptake by plants [91–93]. Endophytes are able to enhance the growth of many plant species with or without concomitant nutrient uptake both directly and indirectly (Table 3.1). However, the impact of endophyte colonization on nutrient uptake in planta can be variable among strains and is considered to be dependent on host species/cultivars, endophyte taxa, and environmental conditions [94]. Although a broad range of endophytes are described with nutrient acquisition traits as reported in this study few endophytes have been studied in detail to conclusively demonstrate the mechanism(s) of nutrient transfer/acquisition of nutrient in planta. The basidiomycete fungal endophyte *Piriformospora indica* has gained substantial interest as a potential growth-promoting agent [95]. *P. indica* may serve as a model system to elucidate the mechanisms of host growth or fitness, as it has the capability in mobilizing plant unavailable P by production of extracellular phosphatases and in translocation of P in plants [95, 96]. *P. indica* stimulates plant growth as well as seed production of many plants and possesses a broad host range specificity [96, 97]. An increasing number of studies on this fungus provide a scientific basis for agricultural application, and also importantly that this fungal endophyte can easily be grown axenically [98–100]. The P acquisition potential of this endophyte was tested in maize, barley, and *Arabidopsis* [51, 101, 102]; further testing its P acquisition potential in a range of plant hosts will validate the mechanistic basis and may identify if P transport or acquisition is a generalized mechanism of nutrient transfer among crops.

Besides *P. indica*, a number of dark septate endophyte (DSE) fungi (the non-clavicipitaceous group of fungi) have been recognized in plants from a range of ecosystems [103]. The DSEs can help increase plant growth by increasing acquisition of plant nutrients such as N and P in certain plants [77–79]. It has been proposed that DSE symbioses, like mycorrhizas, are multifunctional and not limited to nutritional acquisition and host growth response [28]. However the overall functional potential of this class of fungi or its inoculation potential in plants needs to be verified in order to utilize it for plant growth promotion [104]. The clavicipitaceous endophytes are another group of fungi whose role remains elusive, and work is mostly focused on two related genera *Epichloë* and their anamorphic *Neotyphodium* relatives [105]. These fungi are recognized as increasing P nutrients and in certain cases are also known to be involved in other functions such as abiotic stress tolerance [106], remediation of metal contamination [107], and biocontrol activity [105]. This group of fungi mostly inhabit grasses (family Poaceae) and may have potential in plant growth improvement [105].

Bacteria have also been shown to be important with respect to P and Fe acquisition in plants and are involved in P solubilization or mineralization activity and siderophore production as discussed in previous sections in this chapter. The commercialization of endophyte bacteria such as *Gluconacetobacter diazotrophicus* has gained substantial attention, as N-Fix® (Azotic Technologies, UK) or NITROFIX™- AD (AgriLife, India) for biological nitrogen fixation in plants. *G. diazotrophicus* is also recognized for other traits such as phosphate solubilization activity and this function could be synergistic in plants [72]. Strains of *G. diazotrophicus* are isolated in many areas of world and are utilized commercially to enhance plant production [108].

The biofertilizer potential of siderophore-producing endophytes is also of importance in agriculture not only in terms of improving direct Fe availability in plants by siderophore production but also by biocontrol of pathogens indirectly by increasing Fe nutrient status of plant and depriving the pathogen of iron. The siderophore-producing endophyte may also have phytoremediation potential for remediation of contaminated soils [75, 76, 109] and may share functional similarity with phosphate-solubilizing endophytes which have the ability to produce organic acids and similar to siderophore production may assist in the remediation of contaminated soils [75, 76, 110].

The successful manipulation of the plant microbiome has the potential to increase agricultural production [111, 112] and reduce chemical inputs [113–115] and greenhouse gasses [116] which will result in more sustainable agricultural practices. However, this will require a more detailed exploration of the mechanisms involved in P- and Fe-facilitated mobilization by the plant microbiota.

References

1. Neilands JB. Siderophores: structure and function of microbial iron transport compounds. J Biol Chem. 1995;270:26723–6.
2. Chhabra S, Brazil D, Morrissey J, Burke JI, O'Gara FN, Dowling D. Characterization of mineral phosphate solubilization traits from a barley rhizosphere soil functional metagenome. Microbiology. 2013;2:717–24.
3. Sashidhar B, Podile AR. Mineral phosphate solubilization by rhizosphere bacteria and scope for manipulation of the direct oxidation pathway involving glucose dehydrogenase. J Appl Microbiol. 2010;109:1–12.
4. Li G, Kronzucker HJ, Shi W. The response of the root apex in plant adaptation to iron heterogeneity in soil. Front Plant Sci. 2016;7:344.
5. Rausch C, Bucher M. Molecular mechanisms of phosphate transport in plants. Planta. 2002;216:23–37.
6. Vance CP, Uhde-Stone C, Allan DL. Phosphorus acquisition and use: critical adaptations by plants for securing a nonrenewable resource. New Phytol. 2003;157:423–47.
7. Gilroy S, Jones DL. Through form to function: root hair development and nutrient uptake. Trends Plant Sci. 2000;5:56–60.
8. López-Bucio J, Cruz-Ramírez A, Herrera-Estrella L. The role of nutrient availability in regulating root architecture. Curr Opin Plant Biol. 2003;6:280–7.
9. Lynch JP, Brown KM. Topsoil foraging—an architectural adaptation of plants to low phosphorus availability. Plant Soil. 2001;237:225–37.
10. Shen J, Yuan L, Zhang J, Li H, Bai Z, Chen X, Zhang W, Zhang F. Phosphorus dynamics: from soil to plant. Plant Physiol. 2011;156:997–1005.
11. Raghothama KG. Phosphate acquisition. Annu Rev Plant Physiol Plant Mol Biol. 1999;50:665–93.
12. Clark RB, Zeto SK. Mineral acquisition by arbuscular mycorrhizal plants. J Plant Nutr. 2000;23:867–902.
13. Johri AK, Oelmüller R, Dua M, Yadav V, Kumar M, Tuteja N, et al. Fungal association and utilization of phosphate by plants: success, limitations, and future prospects. Front Microbiol. 2015;6:984.
14. Behie SW, Padilla-Guerrero IE, Bidochka MJ. Nutrient transfer to plants by phylogenetically diverse fungi suggests convergent evolutionary strategies in rhizospheric symbionts. Commun Integr Biol. 2013;6:e22321.
15. Compant S, Clement C, Sessitsch A. Plant growth promoting bacteria in the rhizosphere and endosphere of plants: their role, colonization, mechanisms involved and prospects for utilization. Plant Phys Biochem. 2010;42:669–7.
16. Gaiero JR, McCall CA, Thompson KA, Day NJ, Best AS, Dunfield KE. Inside the root microbiome: bacterial root endophytes and plant growth promotion. Am J Bot. 2013;100:1738–50.
17. Richardson AE, Barea J-M, McNeill AM, Prigent-Combaret C. Acquisition of phosphorus and nitrogen in the rhizosphere and plant growth promotion by microorganisms. Plant Soil. 2009;321:305–39.
18. Berendsen RL, Pieterse CMJ, Bakker P. The Rhizosphere microbiome and plant health. Trends Plant Sci. 2012;17:478–6.
19. Berg G, Rybakova G, Koberl M. The plant microbiome explored: implications for experimental botany. J Exp Bot. 2015; doi:10.1093/jxb/erv466.
20. Vandenkoornhuyse P, Quaiser A, Duhamel M, Le Van A, Dufresne A. The importance of the plant microbiome of the plant holobiont. New Phytol. 2015;206:1196–206.
21. Pii Y, Borruso L, Brusetti L, Crecchio C, Cesco S. The interaction between iron nutrition, plant species and soil type shapes the rhizosphere microbiome. Plant Physiol Biochem. 2016;99:39–48.

22. Thijs S, Sillen W, Rineau F, Weyens N, Vangronsveld J. Towards an enhanced understanding of plant-microbiome interactions to improve phytoremediation: engineering the metaorganism. Front Microbiol. 2016;7:341.
23. Mei C, Flinn BS. The use of beneficial microbial endophytes for plant biomass and stress tolerance improvement. Recent Pat Biotechnol. 2010;4:81–95.
24. Richardson AE, Lynch JP, Ryan PR, Delhaize E, Andrew Smith F, Smith SE, Harvey PR, Ryan MH, Veneklass EJ, Lambers H, Oberson A, Culvenor RJ, Simpson RJ. Plant and microbial strategies to improve the phosphorus efficiency of agriculture. Plant Soil. 2011;349:121–56.
25. Spaepen S, Vanderleyden J. Auxin and plant-microbe interactions. Cold Spring Harb Perspect Biol. 2011;3 doi:10.1101/cshperspect.a001438.
26. Brundrett MC. Coevolution of roots and mycorrhizas of land plants. New Phytol. 2002;154:275–304.
27. Corradi N, Brachmann A. Fungal mating in the most widespread plant symbionts? Trends Plant Sci. 2017;22:175–83.
28. Mandyam K, Jumpponen A. Seeking the elusive function of the root-colonising dark septate endophytic fungi. Stud Mycol. 2005;53:173–89.
29. Jumpponen A, Trappe JM. Dark septate endophytes: a review of facultative biotrophic root-colonizing fungi. New Phytol. 1998;140:295–310.
30. Delaux P-M, Varala K, Edger PP, Coruzzi GM, Pires JC, Ané J-M. Comparative phylogenomics uncovers the impact of symbiotic associations on host genome evolution. PLoS Genet. 2014;10:e1004487.
31. Oelmüller R, Sherameti I, Tripathi S, Varma A. Piriformospora indica, a cultivable root endophyte with multiple biotechnological applications. Symbiosis. 2009;49:1–17.
32. Gill SS, Gill R, Trivedi DK, Anjum NA, Sharma KK, Ansari MW, et al. Piriformospora indica: potential and significance in plant stress tolerance. Front Microbiol. 2016;7:332.
33. Dobbelaere S, Okon Y. The plant growth-promoting effect and plant responses. In: Elmerich C, Newton WE, editors. Associative and endophytic nitrogen-fixing bacteria and cyanobacterial associations. Netherlands: Springer; 2007. p. 145–70.
34. Rascovan N, Carbonetto B, Perrig D, Díaz M, Canciani W, Abalo M, Alloati J, González-Anta G, Vazquez MP. Integrated analysis of root microbiomes of soybean and wheat from agricultural fields. Sci Rep. 2016;6:28084.
35. Suman A, Yadav AN, Verma P. Endophytic microbes in crops: diversity and beneficial impact for sustainable agriculture. In: Singh DP, Singh HB, Prabha R, editors. Microbial inoculants in sustainable agricultural productivity. India: Springer; 2016. p. 117–43.
36. Sharma SB, Sayyed RZ, Trivedi MH, Gobi TA. Phosphate solubilizing microbes: sustainable approach for managing phosphorus deficiency in agricultural soils. Spring. 2013;2:587.
37. Ahmed E, Holmström SJM. Siderophores in environmental research: roles and applications. Microb Biotechnol. 2014;7:196–208.
38. Jansson M. Phosphate uptake and utilization by bacteria and algae. Hydrobiologia. 1988;170:177–89.
39. Kobayashi T, Nishizawa NK. Iron uptake, translocation, and regulation in higher plants. Annu Rev Plant Biol. 2012;63:131–52.
40. Schachtman DP, Reid RJ, Ayling SM. Phosphorus uptake by plants: from soil to cell. Plant Physiol. 1998;116:447–53.
41. Otieno N, Lally RD, Kiwanuka S, Lloyd A, Ryan D, Germaine KJ, Dowling DN. Plant growth promotion induced by phosphate solubilizing endophytic Pseudomonas isolates. Front Microbiol. 2015;6:745.
42. Jog R, Pandya M, Nareshkumar G, Rajkumar S. Mechanism of phosphate solubilization and antifungal activity of Streptomyces spp isolated from wheat roots and rhizosphere and their application in improving plant growth. Microbiology. 2014;160:778–88.
43. Chen Y, Fan J-B, Du L, Xu H, Zhang Q-H, He Y-Q. The application of phosphate solubilizing endophyte Pantoea dispersa triggers the microbial community in red acidic soil. Appl Soil Ecol. 2014;84:235–44.

44. Kim KY, Jordan D, McDonald GA. Enterobacter agglomerans, phosphate solubilizing bacteria, and microbial activity in soil: effect of carbon sources. Soil Biol Biochem. 1998;30:995–1003.
45. Mehta P, Walia A, Kakkar N, Shirkot CK. Tricalcium phosphate solubilisation by new endophyte Bacillus methylotrophicus. Acta Physiol Plant. 2014;36:2033–45.
46. Maccheroni W Jr, Azevedo JL. Synthesis and secretion of phosphatases by endophytic isolates of Colletotrichum musae grown under conditions of nutritional starvation. J Gen Appl Microbiol. 1998;44:381–7.
47. Rodríguez H, Fraga R. Phosphate solubilizing bacteria and their role in plant growth promotion. Biotechnol Adv. 1999;17:319–39.
48. Smith FW, Mudge SR, Rae AL, Glassop D. Phosphate transport in plants. Plant Soil. 2003;248:71–83.
49. Pao SS, Paulsen IT, Saier MH Jr. Major facilitator superfamily. Microbiol Mol Biol Rev. 1998;62:1–34.
50. Saier MH, Reizer J. Families and super families of transport proteins common to prokaryotes and eukaryotes. Curr Opin Struct Biol. 1991;1:362–8.
51. Yadav V, Kumar M, Deep DK, Kumar H, Sharma R, Tripathi T, et al. A phosphate transporter from the root endophytic fungus Piriformospora indica plays a role in phosphate transport to the host plant. J Biol Chem. 2010;285:26532–44.
52. Pedersen BP, Kumar H, Waight AB, Risenmay AJ, Roe-Zurz Z, Chau BH, Schlessinger A, Bonomi M, Harries W, Sali A, Johri AK, Stroud RM. Crystal structure of a eukaryotic phosphate transporter. Nature. 2013;496:533–6.
53. Kim SA, Guerinot ML. Mining iron: iron uptake and transport in plants. FEBS Lett. 2007;581:2273–80.
54. Lemanceau P, Expert D, Gaymard F, Bakker PAHM, Briat JF. Role of iron in plant–microbe interactions. Adv Bot Res. 2009;51:491–549.
55. Conte SS, Walker EL. Transporters contributing to iron trafficking in plants. Mol Plant. 2011;4:464–76.
56. Curie C, Panaviene Z, Loulergue C, Dellaporta SL, Briat JF, Walker EL. Maize yellow stripe1 encodes a membrane protein directly involved in Fe(III) uptake. Nature. 2001;409:346–9.
57. Taurian T, Anzuay MS, Ludueña LM, et al. Effects of single and co-inoculation with native phosphate solubilising strain Pantoea sp J49 and the symbiotic nitrogen fixing bacterium Bradyrhizobium sp SEMIA 6144 on peanut (Arachis hypogaea L.) growth. Symbiosis. 2012;59:77–85.
58. Resende MP, Jakoby I. Phosphate solubilization and phytohormone production by endophytic and rhizosphere Trichoderma isolates of guanandi (Calophyllum Brasiliense Cambess). Afr J Microbiol Res. 2014;8:2616–23.
59. Kuklinsky-Sobral J, Araújo WL, Mendes R, et al. Isolation and characterization of soybean-associated bacteria and their potential for plant growth promotion. Environ Microbiol. 2004;6:1244–51.
60. Marlida Y, Delfita R, Adnadi P, et al. Isolation, characterization and production of phytase from endophytic fungus its application for feed. Pak J Nutr. 2010;9:471–4.
61. de Lacerda JRM, da Silva TF, Vollú RE, Marques JM, Seldin L. Generally recognized as safe (GRAS) Lactococcus lactis strains associated with Lippia sidoides Cham. are able to solubilize/mineralize phosphate. Springerplus. 2016;5:828.
62. Lopez BR, Bashan Y, Bacilio M. Endophytic bacteria of Mammillaria fraileana, an endemic rock-colonizing cactus of the southern Sonoran Desert. Arch Microbiol. 2011;193:527–41.
63. Khan AL, Halo BA, Elyassi A, Ali S, Al-Hosni K, Hussain J, et al. Indole acetic acid and ACC deaminase from endophytic bacteria improves the growth of Solarium lycopersicum. Electron J Biotechnol. 2016;19:58–64.
64. Ji SH, Gururani MA, Chun S-C. Isolation and characterization of plant growth promoting endophytic diazotrophic bacteria from Korean rice cultivars. Microbiol Res. 2014;169:83–98.

65. Puente ME, Bashan Y, Li CY, Lebsky VK. Microbial populations and activities in the rhizoplane of rock-weathering desert plants. I. Root colonization and weathering of igneous rocks. Plant Biol. 2004;6:629–42.
66. López-López A, Rogel MA, Ormeño-Orrillo E, et al. Phaseolus vulgaris seed-borne endophytic community with novel bacterial species such as Rhizobium endophyticum sp. nov. Syst Appl Microbiol. 2010;33:322–7.
67. Dinić Z, Ugrinović M, Bosnić P, et al. Solubilization of inorganic phosphate by endophytic Pseudomonas sp. from French bean nodules. Ratarstvo i povrtarstvo. 2014;51:100–5.
68. Jasim B, Jimtha JC, Jyothis M, Radhakrishnan EK. Plant growth promoting potential of endophytic bacteria isolated from Piper nigrum. Plant Growth Regul. 2013;71:1–11.
69. Khan Z, Kandel S, Ramos D, et al. Increased biomass of nursery-grown Douglas-fir seedlings upon inoculation with diazotrophic endophytic consortia. For Trees Livelihoods. 2015;6:3582–93.
70. Crespo JM, Boiardi JL, Luna MF. Mineral phosphate solubilization activity of gluconacetobacter diazotrophicus under P-limitation and plant root environment. Agric Sci. 2011;02:16–22.
71. Hakim SS, Budi SW, Turjaman M. Phosphate solubilizing and antifungal activity of root endophyte isolated from Shorea leprosula Miq. and Shoreal selanica (DC) Blume. J Manajemen Hutan Tropika. 2015;21:138–46.
72. Intorne AC, de Oliveira MVV, Lima ML, da Silva JF, Olivares FL, de Souza Filho GA. Identification and characterization of Gluconacetobacter diazotrophicus mutants defective in the solubilization of phosphorus and zinc. Arch Microbiol. 2009;191:477–83.
73. Hiruma K, Gerlach N, Sacristán S, Nakano RT, Hacquard S, Kracher B, Neumann U, Ramírez D, Bucher M, O'Connell RJ, Schulze-Lefert P. Root endophyte Colletotrichum tofieldiae confers plant fitness benefits that are phosphate status dependent. Cell. 2016;165:464–74.
74. Kumar M, Yadav V, Kumar H, Sharma R, Singh A, Tuteja N, Johri AK. Piriformospora indica enhances plant growth by transferring phosphate. Plant Signal Behav. 2011;6:723–5.
75. Li H-Y, Wei D-Q, Shen M, Zhou Z-P. Endophytes and their role in phytoremediation. Fungal Divers. 2012;54:11–8.
76. Li X, Ren A, Han R, Yin L, Wei M, Gao Y. Endophyte-mediated effects on the growth and physiology of Achnatherum sibiricum are conditional on both N and P availability. PLoS One. 2012;7:e48010.
77. Jumpponen A, Mattson KG, Trappe JM. Mycorrhizal functioning of Phialocephala fortinii with Pinus contorta on glacier forefront soil: interactions with soil nitrogen and organic matter. Mycorrhiza. 1998;7:261–5.
78. Newsham KK. Phialophora graminicola, a dark septate fungus, is a beneficial associate of the grass Vulpia ciliata ssp. ambigua. New Phytol. 1999;144:517–24.
79. Barrow JR, Osuna P. Phosphorus solubilization and uptake by dark septate fungi in fourwing saltbush, Atriplex canescens (Pursh) Nutt. J Arid Environ. 2002;51:449–59.
80. Malinowski DP, Brauer DK, Belesky DP. The endophyte neotyphodium coenophialum affects root morphology of tall fescue grown under phosphorus deficiency. J Agron Crop Sci. 1999;183:53–60.
81. Haselwandter K, Read DJ. The significance of a root-fungus association in two Carex species of high-alpine plant communities. Oecologia. 1982;53:352–4.
82. Gillis M, Kersters K, Hoste B, Janssens D, Kroppenstedt RM, Stephan MP, Teixeira KRS, Dobereiner J, Ley JD. Acetobacter diazotrophicus sp. nov., a nitrogen-fixing acetic acid bacterium associated with sugarcane. Int J Syst Evol Microbiol. 1989;39:361–4.
83. Radzki W, Gutierrez Manero FJ, Algar E, Lucas Garcıa JA, Garcıa-Villaraco A, Ramos Solano B. Bacterial siderophores efficiently provide iron to iron-starved tomato plants in hydroponics culture. Antonie Van Leeuwenhoek. 2013;104:321–30.
84. Koulman A, Lee TV, Fraser K, Johnson L, Arcus V, Lott JS, Rasmussen S, Lane G. Identification of extracellular siderophores and a related peptide from the endophytic fungus Epichloë festucae in culture and endophyte-infected Lolium perenne. Phytochemistry. 2012;75:128–39.

85. Johnson LJ, Koulman A, Christensen M, Lane GA, Fraser K, Forester N, Johnson RD, Bryan GT, Rasmussen S. An extracellular siderophore is required to maintain the mutualistic interaction of Epichloe festucae with Lolium perenne. PLoS Pathog. 2013;9:e1003332. doi:10.1371/journal.ppat.1003332.
86. Rosconi F, Trovero MF, de Souza EM, Fabiano E. Serobactins mediated iron acquisition systems optimize competitive fitness of Herbaspirillum seropedicae inside rice plants. Environ Microbiol. 2016;18(8):2523–33.
87. Maldonado-González MM, Elisabetta Schilirò E, Prieto P, Mercado-Blanco J. Endophytic colonization and biocontrol performance of Pseudomonas fluorescens PICF7 in olive (Olea europaea L.) are determined neither by pyoverdine production nor swimming motility. Environ Microbiol. 2015;17:3139–53.
88. Verma VC, Singh SK, Prakash S. Bio-control and plant growth promotion potential of siderophore producing endophytic Streptomyces from Azadirachta indica A. Juss J Basic Microbiol. 2011;51:550–6.
89. Rungin S, Indananda C, Suttiviriya P, Kruasuwan W, Jaemsaeng R, Thamchaipenet A. Plant growth enhancing effects by a siderophore-producing endophytic streptomycete isolated from a Thai jasmine rice plant (Oryza sativa L. cv. KDML105). Antonie Van Leeuwenhoek. 2012;102:463–72.
90. Morrissey J, Guerinot ML. Iron uptake and transport in plants: the good, the bad, and the ionome. Chem Rev. 2009;109:4553–67.
91. Jilani G, Akram A, Ali RM, Hafeez FY, Shamsi IH, Chaudhry AN, et al. Enhancing crop growth, nutrients availability, economics and beneficial rhizosphere microflora through organic and biofertilizers. Ann Microbiol. 2007;57:177–84.
92. Jin CW, Ye YQ, Zheng SJ. An underground tale: contribution of microbial activity to plant iron acquisition via ecological processes. Ann Bot. 2013;113:7–18.
93. Yazdani M, Bahmanyar MA, Pirdashti H, Esmaili MA. Effect of phosphate solubilization microorganisms (PSM) and plant growth promoting rhizobacteria (PGPR) on yield and yield components of corn (Zea mays L.). Proc World Acad Sci Eng Technol. 2009;37:90–2.
94. Kageyama SA, Mandyam KG, Jumpponen A. Diversity, function and potential applications of the root-associated endophytes. In: Varma PDA, editor. Mycorrhiza. Springer: Berlin, Heidelberg; 2008. p. 29–57.
95. Deshmukh S, Hückelhoven R, Schäfer P, Imani J, Sharma M, Weiss M, et al. The root endophytic fungus Piriformospora indica requires host cell death for proliferation during mutualistic symbiosis with barley. Proc Natl Acad Sci U S A. 2006;103:18450–7.
96. Singh A, Sharma J, Rexer K-H, Varma A. Plant productivity determinants beyond minerals, water and light: Piriformospora indica—a revolutionary plant growth promoting fungus. Curr Sci. 2000;79:1548–54.
97. Singh LP, Gill SS, Tuteja N. Unraveling the role of fungal symbionts in plant abiotic stress tolerance. Plant Signal Behav. 2011;6:175–91.
98. Das A, Kamal S, Shakil NA, Sherameti I, Oelmüller R, Dua M, et al. The root endophyte fungus Piriformospora indica leads to early flowering, higher biomass and altered secondary metabolites of the medicinal plant, Coleus forskohlii. Plant Signal Behav. 2012;7:103–12.
99. Trivedi DK, Verma PK, Srivastava A, Gill SS, Tuteja N. Piriformospora indica: a friend in need is a friend in deed. Res Rev. 2016;5:16–9.
100. Varma A, Sree KS, Arora M, Bajaj R, Prasad R, Kharkwal AC. Functions of novel symbiotic fungus – Piriformospora indica. Proc Indian Natl Sci Acad. 2014;80:429.
101. Achatz B, von Rüden S, Andrade D, Neumann E, Pons-Kühnemann J, Kogel K-H, et al. Root colonization by Piriformospora indica enhances grain yield in barley under diverse nutrient regimes by accelerating plant development. Plant Soil. 2010;333:59–70.
102. Shahollari B, Varma A, Oelmüller R. Expression of a receptor kinase in Arabidopsis roots is stimulated by the basidiomycete Piriformospora indica and the protein accumulates in Triton X-100 insoluble plasma membrane microdomains. J Plant Physiol. 2005;162:945–58.
103. Knapp DG, Pintye A, Kovács GM. The dark side is not fastidious – dark septate endophytic fungi of native and invasive plants of semiarid sandy areas. PLoS One. 2012;7:e32570.

104. Mandyam KG, Jumpponen A. Mutualism–parasitism paradigm synthesized from results of root-endophyte models. Front Microbiol. 2015;5:776.
105. Kuldau G, Bacon C. Clavicipitaceous endophytes: their ability to enhance resistance of grasses to multiple stresses. Biol Control. 2008;46:57–71.
106. Arachevaleta M, Bacon CW, Hoveland CS, Radcliffe D. Effect of the tall fescue endophyte on plant response to environmental stress. Agron J. 1989;81:83.
107. Malinowski DP, Belesky DP. Adaptations of endophyte-infected cool-season grasses to environmental stresses: mechanisms of drought and mineral stress tolerance. Crop Sci. 2000;40:923–40.
108. Saravanan VS, Madhaiyan M, Osborne J, Thangaraju M, Sa TM. Ecological occurrence of Gluconacetobacter diazotrophicus and nitrogen-fixing Acetobacteraceae members: their possible role in plant growth promotion. Microb Ecol. 2007;55:130–40.
109. Deng Z, Cao L. Fungal endophytes and their interactions with plants in phytoremediation: a review. Chemosphere. 2016;168:1100–6.
110. Ahemad M. Phosphate-solubilizing bacteria-assisted phytoremediation of metalliferous soils: a review. 3 Biotech. 2015;5:111.
111. Bakker MG, Manter DK, Sheflin AM, Weir TL, Vivanco JM. Harnessing the rhizosphere microbiome through plant breeding and agricultural management. Plant Soil. 2012;360:1–13.
112. Tkacz A, Poole P. The role of the plant microbiota in productivity. J Exp Bot. 2015;66:2167–75.
113. Adesemoye AO, Torbert HA, Kloepper JW. Plant growth-promoting rhizobacteria allow reduced rate application rates of chemical fertilizers. Microb Ecol. 2009;58:921–9.
114. Germaine KJ, Chhabra S, Song B, Brazil D, Dowling DN. Microbes and sustainable production of biofuel crops: a nitrogen perspective. Biofuels. 2010;1(6):877–88.
115. Kandel SL, Firrincieli A, Joubert PM, Okubara PA, Leston N, McGeorge K, Mugnozza GS, Harfouche A, Kim S-H, Doty SL. An in vitro study of bio-control and plant growth promotion potential of Salicaceae endophytes. Front Microbiol. 2017;8 doi:10.3389/fmicb.2017.00386.
116. Singh BK, Bardgett RD, Smith P, Reay DS. Microorganisms and climate change: terrestrial feedback and mitigation options. Nat Rev Microbiol. 2010;8:779–90.
117. Vendan RT, Yu YJ, Lee SH, Rhee YH. Diversity of endophytic bacteria in ginseng and their potential for plant growth promotion. J Microbiol. 2010;48: 559–565.

Chapter 4
Endophyte Effects on Photosynthesis and Water Use of Plant Hosts: A Meta-Analysis

Hyungmin Rho and Soo-Hyung Kim

4.1 Introduction

Like animals do, plants host microbiomes. These include endophytes, beneficial bacteria, fungi, and yeast forming microbial communities in planta [1]. They are found to enhance growth and fitness of host plants sustaining in harsh environmental conditions by offering them various benefits. These constructive effects include, but are not limited to, biological fixation of the atmospheric di-nitrogen [2, 3], production/promotion/regulation of plant growth hormones [4–6], facilitation of nutrient acquisition [7, 8], and provision of tolerances to abiotic/biotic environmental stressors [9–11]. All of these impacts promote growth and increase fitness of plant hosts and are well documented and summarized throughout the literature [12–14].

Most of the literature on endophytes has emphasized either the symbiotic characteristics/functions of the symbionts in vitro systems at the lab scale or the influences on overall biomass gain at the field scale. Therefore, understanding endophyte effects on plant functional traits has become important to shape our knowledge about symbiotic interactions and to highlight the importance of plant microbiomes [15]. Many studies have uncovered some plant physiological mechanisms affected by endophytes, and the authors provide some mechanistic explanations of endosymbiosis in different contexts. However, the studies are often too species specific or environmental condition specific to provide a consensus of their impacts on functional traits.

The scope of this chapter is centered on photosynthesis and water relations among the plant functional trait differences gained by endophytic symbiosis (Fig. 4.1). Friesen et al. [15] argued that the impacts of endophytes on photosynthetic pathways are implausible (Table 4.1). However, Straub et al. [42] demonstrated that the expression of the genes related to the photosynthetic light reaction

H. Rho • S.-H. Kim (✉)
School of Environmental and Forest Sciences, University of Washington, Seattle, WA, USA
e-mail: soohkim@uw.edu

S.L. Doty (ed.), *Functional Importance of the Plant Microbiome*,
DOI 10.1007/978-3-319-65897-1_4

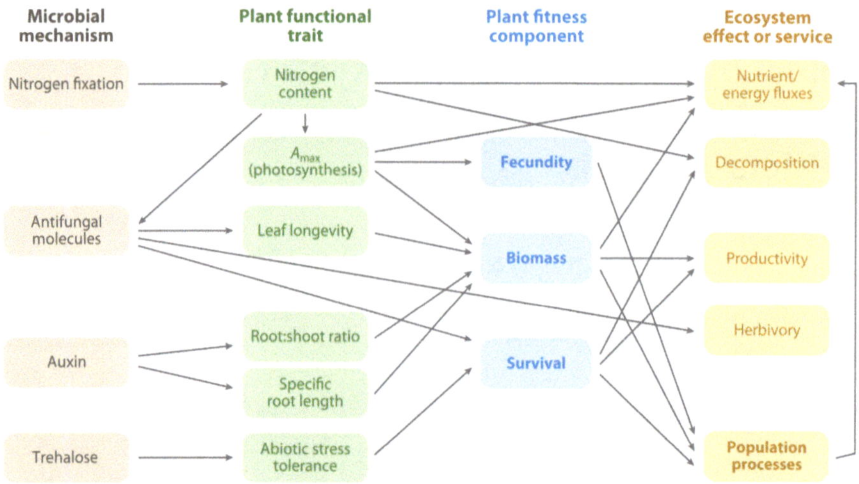

Fig. 4.1 Exemplary interactions between host plants and endophytic microorganisms and their relationships in metabolic and physiological components. Not all connections are described in the diagram. In this chapter, the focus falls on plant functional trait with emphasis on photosynthetic and water use efficiency. Adapted from Friesen et al. [15]

of *Miscanthus sinensis* was upregulated upon a bacterial endophyte inoculation. Ghabooli et al. [43] also showed that photosynthetic enzymes, e.g., Rubisco large/small-chain, chlorophyll a-b-binding proteins, were upregulated by a fungal endophyte inoculation under drought-stress conditions. Indeed, despite some controversial reports, there exists a body of evidence indicating the effects on photosynthetic properties and water relations of plants affected by endophytic symbionts. This chapter reviews these effects of bacterial, fungal, and yeast endophytes on broad photosynthetic characteristics and water relations of plants under a variety of stress-imposing environmental conditions.

4.2 Definitions of Terms and Methods of Analysis

4.2.1 Endophytes

Endophytes can be classified mainly under three categories: bacterial, fungal, and yeast endophytes. They are different domains of microorganisms, but share a common characteristic of living inside plants providing multiple benefits to the host. Most commonly investigated endophytic bacterial genera are *Pseudomonas*, *Bacillus*, *Rhizobium*, *Streptomyces*, and *Staphylococcus* [15]. Fungal species include *Neotyphodium* and *Epichloe* [44] which are the two most studied endophyte strains among all. A study showed that a yeast species—*Rhodotorula*—can form an endophytic mutualism with various plant hosts [45]. One noteworthy approach is

Table 4.1 Summary of studies used in meta-analysis

Reference	Type of endophytes	Genus of endophytes	Species of host plants	Leaf physiology parameters	Number of data sets	Stressor	Experimental environment
Ali et al. [16]	Bacteria	*Pseudomonas*	*Solanum lycopersicum*	Chl, Flr	8	Salinity	Greenhouse
Barnawal et al. [4]	Bacteria	*Brachybacterium*	*Chlorophytum*	Chl	1	Salinity	Greenhouse
Bu et al. [17]	Fungi	*Suaeda*	*Oryza sativa*	Chl, A, g_s, E, WUE	4	Salinity	Chamber
Gagné-Bourque et al. [18]	Bacteria	*Bacillus*	*Phleum pratense*	A, g_s	2	Drought	Chamber
Ghimire and Craven [19]	Fungi	*Sebacina*	*Panicum virgatum*	Chl, Flr	3	Drought	Greenhouse
Gond et al. [20]	Bacteria	*Pantoea*	*Zea mays*	Flr	1	Salinity	Lab
Khan et al. [21]	Fungi	*Exophiala*	*Cucumis sativus*	Chl, Flr	1	Heat	Chamber
Khan and Lee [22]	Fungi	*Penicillium*	*Glycine max*	Chl, Flr	1	Metal	Chamber
Khan et al. [23]	Fungi	*Metarhizium*	*Glycine max*	Chl, Flr, A, g_s, E, WUE	2	Salinity	Chamber
Khan et al. [24]	Fungi	*Penicillium*	*Glycine max*	Chl, A, g_s, E, WUE	2	Salinity	Chamber
Khan et al. [9]	Bacteria and yeast	Consortia of *Acinetobacter, Burkholderia, Curtobacterium, Pseudomonas, Rahnella, Rhizobium, Rhodotorula,* and *Sphingomonas*	*Populus deltoides x P. nigra*	Chl, Flr, g_s	1	Drought	Greenhouse

(continued)

Table 4.1 (continued)

Reference	Type of endophytes	Genus of endophytes	Species of host plants	Leaf physiology parameters	Number of data sets	Stressor	Experimental environment
Knoth et al. [25]	Bacteria	Consortia of *Acinetobacter, Burkholderia, Herbaspirillum, Pseudomonas, Rahnella, Rhizobium, Rhodotorula, and Sphingomonas*	*Zea mays*	Chl, Flr, A, g_s, E, WUE	6	Nutrient	Field
Knoth et al. [3]	Bacteria and yeast	Consortia of *Acinetobacter, Burkholderia, Enterobacter, Herbaspirillum, Pseudomonas, Rahnella, Rhizobium, Rhodotorula, and Sphingomonas*	*Populus trichocarpa* clone Nisqually-1	Chl, Flr, A, g_s, E, WUE	3	Nutrient	Field
Li et al. [26]	Fungi	*Neotyphodium*	*Achnatherum sibiricum*	Flr	4	Nutrient	Chamber
Li et al. [27]	Fungi	*Suaeda*	*Oryza sativa*	Chl, A, g_s, E, WUE	3	Metal	Chamber
Madhaiyan et al. [28]	Bacteria	*Enterobacter*	*Jatropha curas*	Chl	1	Nutrient	Greenhouse
Morse et al. [29]	Fungi	*Neotyphodium*	*Festuca arizonica*	Flr, A, g_s, WUE	1	Drought	Field
Naveed et al. [30]	Bacteria	*Burkholderia*	*Zea mays*	Chl, A, g_s, E, WUE	4	Drought	Greenhouse
Newman et al. [31]	Fungi	*Neotyphodium*	*Festuca arundinacea*	Flr, A	1	Nutrient	Field
Patel and Saraf [32]	Bacteria	*Enterobacter*	*Jatropha curas*	Chl, Flr	3	Salinity	Lab

Rogers et al. [33]	Bacteria	*Enterobacter*	*Populus deltoides x P. nigra*	A, g_s, WUE	1	–	Field
Rojas-Tapias et al. [34]	Bacteria	*Azotobacter*	*Zea mays*	Chl, Flr	4	Salinity	Chamber
Shahabivand et al. [35]	Fungi	*Piriformospora*	*Triticum aestivum*	Chl, Flr, A	3	Metal	Chamber
Shukla et al. [36]	Fungi	*Trichoderma*	*Oryza sativa*	Chl, A, g_s, WUE	5	Drought	Lab
Vahabi et al. [37]	Fungi	*Piriformospora*	*Arabidopsis thaliana*	Flr	2	Shading	Chamber
Waqas et al. [5]	Fungi	*Penicillium*	*Oryza sativa*	Chl, Flr	8	–	Chamber
Woodward et al. [38]	Fungi	*Fusarium*	*Oryza sativa*	A	2	–	Lab
Yang et al. [39]	Fungi	*Phomopsis*	*Oryza sativa*	Chl	2	Nutrient	Chamber
Zarea et al. [40]	Fungi	*Piriformospora*	*Triticum aestivum*	Chl	6	Salinity	Greenhouse
Zhang and Nan [41]	Fungi	*Neotyphodium*	*Elymus dahuricus*	Chl	1	Drought	Chamber

blending multiple strains of bacterial or yeast endophytes together in hopes of stacking the functional benefits of the endophytes when mixed [3, 9, 23, 25, 46, 47]. The mixture is called a consortium and it is considered in our meta-analysis. The articles referred in this chapter contain all these categories of endophytes (Table 4.1).

4.2.2 Photosynthetic Efficiency

Photosynthesis can be defined broadly as CO_2 uptake or O_2 production or specifically as CO_2 assimilation rate, photosynthetic light harvesting efficiency, chlorophyll content, or biochemical properties of the Calvin-Benson cycle (Rubisco content/activity). The specific terms are used to represent photosynthetic capacity and responses of leaf physiology to environmental cues. Sometimes photosynthesis as a net assimilation rate is expressed by certain biomass terms [48], but we restrain our discussion here to leaf-scale photosynthetic capacity. In relation to this, there are two key processes of photosynthesis in leaf physiology: the light reactions and the dark reactions or carbon reactions.

The light reactions are related to the light harvesting process which takes place in the lamellae sides of chloroplasts in mesophyll cells. Chlorophyll concentration and its efficiency are important factors in determining the efficiency of the first light harvesting aspect of photosynthesis. The most frequently used parameters are chlorophyll content (Chl) and photochemical efficiency (Flr). Chlorophyll content can represent the efficiency of the light harvesting process, while Flr can represent the efficiency of electron transport during the following photochemical reactions in between photosystem II (PSII) and photosystem I (PSI). Chlorophyll can be quantified with various experimental methods, from in vitro quantification [49] to in vivo estimation [50]. These have been developed and widely used in plant sciences. Both methods were considered as a representative for Chl in our analysis. Photochemical efficiency refers to the rate of the light energy that is processed to the available form of biochemical energy—ATP and NADPH—in further photosynthetic CO_2 assimilation processes. It can be measured by a chlorophyll fluorescence technique which is currently broadly implemented in plant stress physiology [51]. Especially, F_v/F_m, defined as maximum quantum efficiency of PSII photochemistry, was the focus of our meta-analysis to evaluate photochemical efficiency.

The carbon reactions, also known as dark reactions, on the other hand, are associated with the CO_2 assimilation process on the stroma sides of chloroplasts in mesophyll cells/bundle sheath cells in C_3/C_4 plants [52]. It begins with the uptake of atmospheric CO_2 from the leaf surfaces via stomata. By diffusion through the intercellular structures of the leaf, CO_2 molecules encounter a series of resistances that limit the supply of CO_2 to the Calvin-Benson cycle at the site of carboxylation. Once CO_2 is captured, it is staged to the carboxylation process where Rubisco catalyzes C fixation. All the enzymes related to operating the Calvin-Benson cycle work to incorporate CO_2 molecules to the three C skeletons. The efficiency of the carboxylation process can be defined by how fast this CO_2 assimilation process occurs by the activities of the associated enzymes. The rate of net CO_2 assimilation (*A*) now can be measured easily in vivo by a commercial gas exchange system equipped with infrared gas analyzers (IRGAs) [53].

4.2.3 Stomatal Conductance and Transpiration

Stomata—the interface of gas exchange on the leaf surfaces—play a crucial role in both photosynthetic CO_2 uptake and coupled H_2O loss of plants [54]. As stomata are the location where CO_2 and H_2O are exchanged mostly during the daytime, it is important to evaluate stomatal reactions to connect photosynthesis with water relations of plants.

Stomata open and close in response to many environmental cues, such as vapor pressure deficit (VPD) [55], atmospheric CO_2 concentration [56], water availability from the rhizosphere [57], and pathogenic attacks [58]. The opening/closure—the aperture—of stomata can be determined by either microscopic observations [59] or in vivo gas exchange activity measurements [60]. To represent the actual amount of water release and CO_2 uptake during the stomatal actions, instantaneous stomatal conductance (g_s) is used as a unit of stomatal activity. Stomatal conductance has become the most commonly used leaf physiology parameter. It also finds significance in linking photosynthesis to water relations in various model approaches [61] and is a reciprocal of the CO_2 diffusional resistance by the movement of guard cells. It can be expressed in conductance of water vapor or carbon dioxide. Stomatal conductance of water vapor can be determined by measuring the rate or the velocity of water vapor diffusion in vivo.

Transpiration (*E*) represents how much water is lost throughout stomata in response to VPD between the atmospheric air and the leaf surfaces. It is a parameter that expresses gas exchanges and water relations of plants. Along with g_s, temperature and relative humidity (RH) in the ambient air, which in turn influence VPD, come into play in determining *E*. Many plant species lose water through stomata and this is thought to occur only in the daytime as the transpiration demand increases when the air temperature is elevated while RH decreases. However, Snyder et al. [62] reported that some species had positive g_s and *E* at night in an arid environment. This suggests that some plants actively transpire under water-limited conditions as a survival strategy. It highlights the importance of g_s and *E* in water management of plants.

In many cases, g_s and *E* are estimated or measured by a commercial IRGA system together with other leaf photosynthetic and gas exchange parameters, for instance, *A* and F_v/F_m.

4.2.4 Water Use Efficiency

As water becomes scarce in irrigated agricultural crop lands, now it is well recognized as one of the key resources for crop yield [63]. Recent efforts in the current challenging climate conditions focus on how to increase the productivity of crops with limited water [64]. As a consequence, water use efficiency (WUE) stands great importance in agriculture.

Plant WUE is cataloged into three measures: intrinsic WUE, extrinsic WUE, and WUE of productivity. The former two are leaf physiological parameters, whereas the latter one is a whole-plant physiological parameter. To understand these parameters, we revisit aforementioned leaf gas exchange parameters: A, g_s, and E. The basic idea of all these WUE is to express how much CO_2 is incorporated into plant biomass with a given amount of water consumed during the process [65].

Intrinsic WUE refers to the WUE determined by inherent leaf physiological processes. It is calculated by "A/g_s." Extrinsic WUE refers to the WUE externally determined by external surrounding factors, i.e., air temperature, RH, and VPD that would affect E of plants. It is calculated by "A/E." These are short-term measures of WUE. A gas exchange machine can compute these WUE instantaneously on the leaf being measured.

In contrast, WUE of productivity is a long-term measure of WUE. It is calculated by "total biomass gain/water consumption" over time. WUE of productivity could be an accumulated result of intrinsic/extrinsic WUE. One can quantify WUE of productivity by measuring the differences in biomass at between initial and harvest stages, together with the records of water consumption.

4.2.5 *Meta-Analysis*

Several studies reviewed plant-endophyte interactions with respect to plant growth and stress [12, 14, 66, 67]. However, few of these review articles addressed eco-physiological processes involving photosynthesis and water relations of the host plants. There is also limited analytic review for generalized photosynthetic and water use responses of endophyte symbiotic plants with few exceptions [68, 69].

Meta-analysis is a means to quantitatively evaluate an overall effect of a certain treatment on different subjects reported in various studies [70]. In general, it is capable of addressing the heterogeneity of studies used in the analysis. Compared to conventional literature reviews, it has an advantage of enabling researchers to make statistical conclusions on research questions from different studies. One can design a meta-analysis starting with refining keywords to search articles that meet the investigators' criteria. Collected data are processed to estimate "effect size" which represents the effectiveness of a treatment on response variable(s). If the variances of the effect sizes, usually presented by 95% confidence intervals (CI), do not overlap the zero effects on the scale, then one can conclude that the effects are statistically significant on increasing/decreasing the reported response variable(s).

In the present meta-analysis, research questions we posed were the following: (1) Do endophytes improve photosynthetic and water use efficiency of host plants? (2) Do the endophyte effects differ when the hosts are under stress in comparison to non-stress conditions? "Endophyte inoculation" is the treatment to be estimated for its effect size and the leaf physiology parameters mentioned above (i.e., Chl, Flr, A,

g_s, E, and WUE) are the response variables to be analyzed. We gathered a total of 30 articles from the SCOPUS (https://www.scopus.com/) database and statistically meta-analyzed 93 data sets extracted from the papers. Details about the articles used in this chapter are provided in Table 4.1.

4.3 Synthesis of Information

A total of 37 articles were gathered, but only 30 were used due to lack of necessary information for a meta-analysis in the other seven articles. Since some articles reported results from multiple strains or experiments, we procured 86 data sets from the selected articles. Only three studies did not conduct endophyte experiments under abiotic/biotic stress conditions. Most other studies tested the endophyte effects under stress conditions compared to non-stressed controls. Most of the plant species used in the studies were important agricultural crops, e.g., rice, corn, wheat, tomatoes, and soybeans. This suggests that there is a consensus of attempting to use endophyte symbioses as growth promotion and stress mitigation means for crops particularly in light of climate change, sustainability, and resource use efficiencies in agriculture [66].

We analyzed 11/18/3 for bacterial/fungal/yeast endophytes used as a single strain or a consortium in the analysis. These different types of endophyte inoculants were introduced to experimental plants in various ways: coating seed surfaces with inocula or soaking seeds with inocula before sowing [19, 20, 32], applying inocula to seedlings grown in aseptic conditions [34], soaking stem cuttings in inocula [33], or directly drenching inocula to the soils where the plant materials were prepared [9]. In some cases, plants were infected by vertically transmitted endophytes from a parental generation while non-infected plants served as controls [29, 71], which is only found in fungal endophyte cases (Classes 1 and 2 fungal endophytes as discussed in [72]). Horizontally transmitted endophytes from neighboring plants were also found in one study [73].

Most of the studies were carried out in well-controlled environments. Except four studies, all were lab, chamber, or greenhouse experiments. This likely becomes a potential drawback of the endophyte research which hinders us from making strong conclusions about their effects at the field scales.

4.3.1 Improvement of Photosynthetic Efficiency by Endophytes: Both the Light and the Dark Reactions

Endophyte inoculation effects under different experimental conditions were found highly significant on increasing photosynthetic efficiency represented by Chl, Flr, and A (Figs. 4.2, A1, A2 and A3). In addition, the effects on the increases were more pronounced under stressed conditions than under non-stressed conditions except Chl of plants. The results suggest that both photosynthetic light reactions and carbon reactions were effectively enhanced by symbiotic associations.

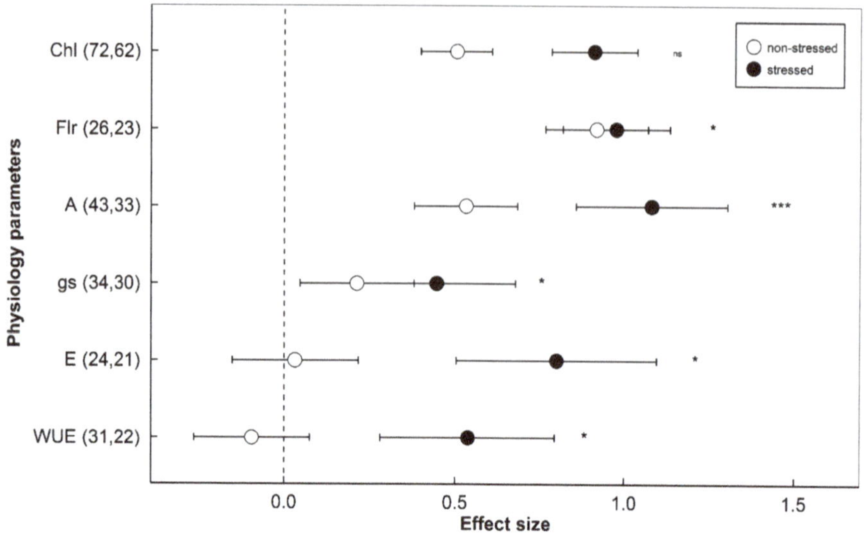

Fig. 4.2 Overall effect sizes of various endophytes on host plants' photosynthetic and water relation parameters. A total of 84 studies are used in the meta-analysis. Chl, Flr, A, gs, E, and WUE refer to chlorophyll content, photochemical efficiency, net CO_2 assimilation rate, stomatal conductance to water, transpiration rate, and intrinsic water use efficiency defined by A/gs. The numbers in parentheses mean the number of data sets used in the meta-analysis for each parameter (under non-stressed, stressed conditions). Open and closed circles show the mean overall effect sizes of endophyte inoculation on the physiological parameters under non-stressed and stressed conditions. Error bars indicate ±95% confidence intervals of the means calculated by the Hedge's method [70]. If the error bars overlap the dotted line (effect size = 0.0), the effect is considered nonsignificant. Significance codes on the right present paired *t*-test results of the differences of the effect sizes between non-stressed and stressed host plants (ns, nonsignificant; * and *** significant at $P < 0.05$ and 0.001 levels). Detailed breakdowns of the effect sizes of individual studies on each physiological parameter are provided in supplementary figures (Figs. A1–A6)

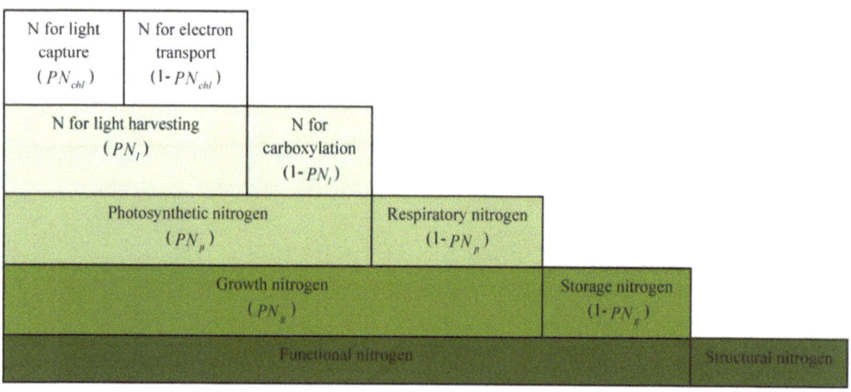

Fig. 4.3 Nitrogen is an essential nutrient in plant metabolism. Up to 40% of growth nitrogen is incorporated to photosynthetic machinery components enclosing light harvesting enzymes and CO_2 assimilation-associated enzymes, e.g., Rubisco, in the Calvin-Benson cycle. This determines photosynthetic efficiency under different nitrogen availability. Adapted from Xu et al. [75]

The increase in photosynthetic light reaction activities characterized by Chl and Flr was one of the most common leaf physiological responses from the studies investigated. Especially, due to the simplicity of the assessment, Chl measured by a handheld chlorophyll meter was the most frequently used leaf physiology parameter ($n = 72$). The observed results were likely to be derived from better nitrogen (N) availability by either symbiotic N fixation [3, 25] or improved N nutrient uptake and assimilation [76]. Chl is a N-rich molecule that consists of up to 75% of leaf N [77]. Chl has a strong positive correlation with leaf N level in a variety of plant species [78]. Scaling down to a molecular level, the upregulation of associated gene expression supported this hypothesis [38, 79]. Chl has a positive correlation with Flr [80]. The increases in a Chl content often lead to increases in Flr.

Separately from the light reactions, the dark reactions are also affected by endophyte symbiosis in a positive direction. There are multiple factors that determine the rate of gas exchange, including stomatal, mesophyll, or biochemical limitations [81]. Among those, considering the integral part of the CO_2 assimilation process, alleviation of biochemical limitation, represented by upregulation of Rubisco-engaged genes, is crucial evidence for the improvement [43, 82]. This is also related to the available N level in leaves as approximately 50% and 20% of N is allocated to Rubisco biosynthesis in C_3 and C_4 plants, respectively [83]. Therefore, as Friesen et al. [15] discussed, N allocation and metabolism are the core parameters resulting in these beneficial effects on photosynthetic efficiency.

However, in some cases, no apparent changes to photosynthetic components of the symbiotic plants were observed [25, 33]. Several reports found different effects on photosynthesis at various stages of growth [25, 29]. This explains the complexity of the endophyte study and increases the uncertainty of implications. Costa Pinto et al. [84] found impairment of the light reactions due to fungal endophyte infection in banana and maize plants in their early stage of growth (30–45 days after seeding). Similarly, Belesky et al. [85] observed significantly lower A in endophyte-infected tall fescue. Interesting findings are of both Belesky et al. [85] and Rogers et al. [33] who reported increases in total biomass accumulation by endophytes though photosynthetic capacity was decreased and not changed. The former paper found the explanation from the increases in tiller numbers and the latter paper did from the increases in leaf area of the plants. It seems that the endophyte-infected plants opted to invest carbon to producing more source organs rather to boosting productivity of sink tissues. Still, the mechanisms underlying these growth dynamics are largely unknown, requiring more research to fully benefit from the symbiotic interactions and to maximize the potential use of this biological adaptation strategy.

4.3.2 Importance of Leaf N Level in Photosynthetic Efficiency and Contribution of Biological N Fixation

Nitrogen is the most important element in plant nutrition and metabolism other than C, H, and O. Approximately 1.5% of the total biomass is composed of N. It is the most abundant macronutrient obtained from the soil [52]. Plants actively absorb N mostly from the root systems in the form of NO_3^- or NH_4^+. Plants then transport N to shoots or leaves where N assimilation occurs to convert them into bioavailable forms (amino acids) by a series of enzymatic responses. These available N sources are then delivered to the photosynthetically active leaf tissues—newer leaves rather than older leaves [74]. Plants invest N into photosynthetic machineries in response to the environmental and other nutrient conditions [75]. Either the light harvesting process, the electron transport process in PSII, or the CO_2 assimilation process can be a target of the investment. Note that each process relates to Chl, Flr, or *A* in the analysis. Evans and Terashima [86] demonstrated that these three parameters (shown as Chl, PSII activity, and Rubisco activity in the original article) were most strongly correlated to leaf N content than any other processes of photosynthetic machineries.

Nitrogen, therefore, alters leaf-related plant functional traits in a significant way through this allocation adjustment between photosynthetic functions [15]. It is straightforward to connect plant N to photosynthetic efficiency as it consists of a substantial portion of growth N. It is not surprising that some studies postulated that the improvement of photosynthesis and further overall biomass by endophyte inoculation was due to increased N availability from biological N fixation (BNF) by the symbionts [3].

As a consequence, to capitalize on these photosynthetic benefits, it is imperative to assess BNF activities in different strains of endophytic microbes, both in vitro and in vivo on a smaller scale. Doty et al. [2] demonstrated the variable capacity of ^{15}N incorporation in cuttings of wild poplar plants, and suggested that specific strains may be required for the high N-fixation activity. Identification of these key strains from the symbiotic microorganism pools is required to develop better combinations of endophyte consortia to augment the extent of the benefits. On a larger scale, it will be required to evaluate to what extent endophytic BNF gives benefits through improving nitrogen use efficiency (NUE) in host plants. Endophyte- and other associative bacteria-engaged BNF is estimated to be 0.5, <4, or <14 Tg/year in sugar cane, non-legume crop, and grazing croplands, respectively [87]. This contributes to 25–35% of total BNF of 50–70 Tg/year in agricultural croplands. Alternation of N uptake efficiency (NUpE), N utilization efficiency (NUtE), or photosynthetic N use efficiency (pNUE) by endophyte symbioses should be examined to compare this sustainable method with other recent approaches to enhance photosynthetic capacity [88–91]. In the case of legume-rhizobium symbiosis, this benefit of BNF will be even more substantial in the future climate conditions with elevated atmospheric CO_2 concentrations [92].

4.3.3 On a Cost-Benefit Approach of Photosynthates

Despite convincing evidence of the photosynthetic improvements, one should also consider the cost of endophyte symbiosis. That is, the additional carbohydrates fixed in the host plants through an increase in photosynthetic capacity or in leaf area facilitated by the endophytes might not be allocated directly to sink tissues where new biomass synthesis occurs but drained to feed and harbor endophytes. Mutualists, like pathogenic microorganisms, traffic the sugar transport through the phloem stream. Likewise, it is a well-established finding that symbionts demand the cost for their benefits to the hosts in the form of sucrose or inorganic acids [93].

To date, carbohydrate costs of endosymbiosis have not been actively discussed compared to the other types of plant-microbe symbioses such as legume-rhizobia and plant-mycorrhizal associations. Researchers have used cost-benefit analysis to quantify carbon cost of these other symbiotic microbes. For instance, 20% of photoassimilates were used by mycorrhizae [94], but it varied from 4% to 20% specifically in arbuscular mycorrhizal symbiosis [95]. The majority (83%) of this photosynthate use was the respiratory loss by the fungi themselves and/or the affected increases in plant root respiration [96]. Likewise, symbiotic N_2-fixing rhizobacteria cost the plants about 14–25% of photosynthates [65, 97]. In certain cases, microbial strains may cheat the hosts to drain more carbon-based food sources when the plants have multiple symbiotic partners including those cheaters [98, 99].

Therefore, the increases in A should not be directly interpreted as the increases in overall biomass production or yield of plants as A is a proxy of instantaneous CO_2 assimilation determined by simultaneous gas exchanges on the leaf surfaces. The symbionts may lead to an increased production of photosynthates as a means to mine their own investments towards the hosts. Figure 4.2 shows that there were 50–100% (corresponding to 0.5–1.0 effect size on the scale) increases in A upon endophyte associations; however, a significant part of these increases could be reduced by the C drain from the symbionts. Assuming that they serve as active C sinks in plants, it would be interesting to test how strong their sink strength is. Providing classical cost-benefit analysis results would also be informative for pinpointing the actual C gain over loss of the plants.

4.3.4 *Varying Endophyte Effects on Water Relations: Stomatal Control in Different Contexts*

Compared to significant improvements of photosynthetic efficiency, the effects on water relations and use efficiency were more variable throughout the literature we analyzed (Figs. 4.2, A4, A5 and A6). Stomatal conductance was increased by endophyte inoculation under both non-stressed and stressed conditions. Transpiration rate and WUE were not changed under non-stressed conditions while they were increased by endophyte inoculation only under stressed conditions.

Overall, g_s increased with endophytes and this stomatal opening could explain the increases in A by allowing more atmospheric CO_2 to diffuse into the leaf inside. This helps plants produce more carbohydrates at the risk of losing H_2O on the leaf surfaces. For example, Shukla et al. [36] conducted a study with rice and five strains of fungal endophytes under drought-stress conditions in which study endophyte inoculation delayed the onset of drought and induced reductions in photosynthesis and other gas exchange parameters—g_s together with *A* and Chl.

On the contrary, Malinowski and Belesky [100] reported several cases of the opposing endophyte effects on g_s. Fungal endophyte infection decreased g_s of grasses that were subjected to drought-stress conditions. These strains include *Neotyphodium*, *Phialophora*, and *Acremonium* families [85, 101–105]. They hypothesized that endophyte inoculation could induce stress metabolism of hosts, leading to preconditioning the hosts to any other biotic stresses. These earlier observations in the fungal endophyte research suggested abscisic acids as a possible hormonal signal to mediate g_s responses, but few experimental approaches have been made to test the hypothesis. Also, increased secondary metabolites [10, 106] in symbiotic plants without stressors support this hypothesis. A recent bacterial and yeast endophyte study [9] showed similar mechanistic arrays of evidence; the rapid stomatal closure to drought stress promoted the tolerance of the host plants. This seemed to be related to the hormonal status of inoculated plants as the strains used in the study were known to produce stress-stimulated jasmonic acids, salicylic acids, and abscisic acids. Also, the inoculated hybrid poplar plants showed a reduced production of reactive oxygen species, which is a strong indicator of enhanced stress responses.

This adaptive strategy with endophytes contrasts to other symbiotic styles. A good example is mycorrhizal symbiosis. Mycorrhizal fungi stimulate stomatal opening in such a way to increase photosynthetic responses of the hosts under stressful conditions [107]. They tend to provide more H_2O to the hosts by increasing surface areas of root systems rather than conserving more H_2O by regulating stomatal opening [108]. Even the directions of symbiotic mechanisms on the same functional traits are opposite to each other, both mutualisms provide the hosts with a better chance to tolerate water deficits.

Some researchers discovered osmotic adjustment as the reason for the improved drought-stress tolerance with regard to efficient water use [104]. This mode of action is different from the stomatal closure theory. Rather, endophyte-infected plants show higher g_s than uninfected plants and this causes the plants to increase the photosynthetic gas exchange. In turn, this increases carbohydrates available for osmotic adjustment as a conditioning tool to defend against the threat of drought. Decreases in complex sugars (e.g., fructans) and subsequent increases in simple sugars suggest this mechanistic adaptation.

4.3.5 Alternative Views on Varying Endophyte Effects on WUE

The promotion of rapid stomatal responses in reaction to the environmental changes likely leads to the decreases in water loss by transpiration, and in turn to the increases in WUE over time. This interpretation is in line with Franks et al. [109] genetic engineering approach to increase WUE of plants. They manipulated the genes involved in stomatal development to decrease stomatal density, specifically by knocking out *epf2* genes and subsequent drops in g_s over the course of drought stresses. They concluded that the reduction of water loss while maintaining photosynthetic activity was the key to improve WUE.

Interestingly, a number of articles analyzed in the present study reported no changes to water relations with certain strains of endophytes inoculated to some plants [3, 25, 33, 36]. Furthermore, there are contrasting reports about endophyte effects on WUE found in the literature. For example, lower WUE was measured in endophyte-infected perennial ryegrass in response to drought [113] as opposed to higher WUE reported in salinity stress [17].

Changes in WUE induced by endophytes also seem to be affected by water availability of the soils. Morse et al. [29] observed decreased WUE in endophyte-infected *Festuca arizonica* under well-watered conditions while contrasting increased WUE under water-limiting conditions. The authors suggested that stomatal closure might help plants withhold water in leaves, leading to conserve soil moisture when drought began. With less use of water, the symbiotic plants could conserve soil moisture during the drought. Subsequently, as the drought became severe, the conserved water allowed the plants to maintain photosynthetic gas exchanges—A and g_s—that probably increased WUE over a long-term period. Bae et al. [82] also demonstrated similar changes in the gas exchange properties of *Theobroma cacao* plants inoculated with *Trichoderma* spp. under drought conditions.

Nonetheless, it is noteworthy that WUE used in this chapter is a measure of instant responses of gas exchanges. It does not necessarily lead to increases in long-term, biomass-based WUE. There are other factors to be considered in a long-term measure; for example, biomass production is caused not only by the amount of photoassimilates, but also by hormonal impacts. Auxin is one of the major phytohormones that is known to stimulate growth responses of plants. Many of the endophyte strains reported were found to produce microbial IAA (indole-3-acetic acids, a naturally occurring auxin) or thought to promote the production of endogenous auxins [4, 23, 110, 111].

4.3.6 Other Considerations on Endophyte Effects and Dynamics of the Associations

Some articles demonstrate that there were not significant gains in physiological parameters in response to endophyte inoculations. Rather, they pointed out the changes in plant fitness component (i.e., biomass) by alteration of endogenous hormonal balance [33].

The same endophyte does not necessarily produce the same treatment effects over time. For instance, Knoth et al. [25] showed different endophyte effects on plant physiology at different life stages of corn plants suggesting that seasonal variation should also be considered.

The influence of environmental stresses is another consideration. The overall effect sizes of endophytes on all the parameters analyzed here were larger under stressed conditions despite different types of stress factors with a varying range of the intensity of the stressors (Table 4.1). As an example, Bu et al. [17] investigated effects of fungal endophytes, *Suaeda salsa*, on photosynthetic ability of *Oryza sativa* under five Na_2CO_3 stress conditions (0, 5, 10, 15, and 20 mM Na_2CO_3). They found that A and E increased in inoculated plants compared to non-inoculated controls under all Na_2CO_3 stress levels. However, WUE, g_s, and Flr in inoculated plants showed stress intensity-dependent responses; they were increased under high Na_2CO_3, while there was no significant difference under low Na_2CO_3. These results suggested that the fungal endophyte effects on improvement of plant performance represented by the photosynthetic parameters are dependent on Na_2CO_3 stress level.

Surrounding environmental conditions also affect the benefits of symbiosis in different ways. Davitt et al. [112] demonstrated the context dependency of symbiosis by differing light intensity. Endophyte effects have plasticity and plants are also flexible in their response to symbiotic stimulation of physiological traits. From an experimenter's point of view, this could bring a challenge. Assessment of a certain strain/consortium can be difficult and reproducibility can be problematic. On top of that, endophyte effects can be species specific. Interaction effects should also be considered when endophytes are used as a factor of the experiments that have multiple treatment factors employed. Morse et al. [29] showed the interaction effects of fungal endophyte inoculation and drought stress treatment on A, g_s, E, and WUE over time. Saikkonen et al. [44] reviewed variable endophyte-plant interactions and pointed out that the interactions could range from antagonistic to mutualistic. The direction of the relationships could be determined by mode of transmission, pattern of infections, and life span of the host plants. This shows dynamic responses of the host plants to endosymbionts and poses challenges to endophyte research.

4.4 Conclusions

Regardless of types or strains, overall, endophytes were shown to enhance photosynthetic capacities of various host plants. They also improved water relations of host plants through different physiological pathways. A minority of reports pointed to negative endophyte effects on both aspects of plant physiology. Indeed, how to explicitly explain these varying effects remains challenging.

Increased N availability through direct provision or indirect stimulation by endophytes seems to be the main reason for the photosynthetic improvement in most of the studies. Tighter stomatal control or osmotic regulation, and further water management induced by endophytes, seems to be the main reason for the better water use efficiency.

However, based on the contexts of experimental settings or treatments, mechanistic explanations could differ. Future endophyte physiology studies may focus on detailed biochemical and molecular level examinations to support lower level changes in the signaling cascades that shape plant functional traits. To facilitate this process, constructing endophyte mutants lacking genes encoding some functional in vitro characteristics, plant mutants lacking genes involved in some functional traits or a combination of them will be required to conduct functional trait studies.

The benefits of endophytes in plant physiological functions are clear, but the costs are still uncharacterized. What are the C costs of the endosymbiosis? How do plants govern these interactions? What environmental cue/condition can encourage endophyte-plant associations and benefit the two the most? All these questions need to be answered for better understanding of the eco-physiology of plant-endophyte interactions.

List of Papers Used in the Meta-Analysis. Find Details in Table 4.1 Refs. [3–5, 9, 16–22, 24–41, 45].

Appendix: Supplemental Figures

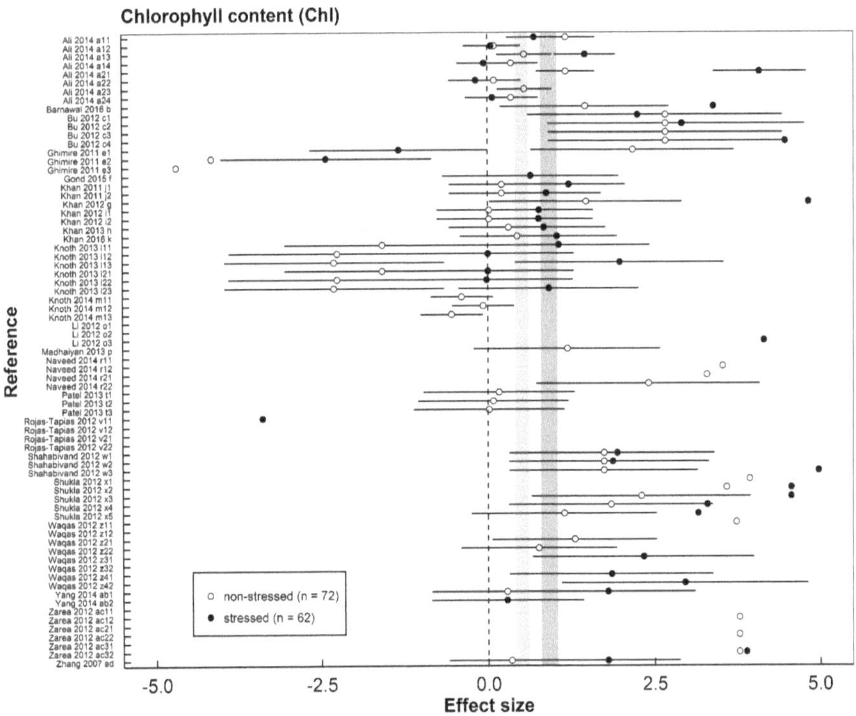

Fig. A1 Effect sizes of various endophytes on host plants' chlorophyll content (Chl). *Open* and *closed circles* indicate the mean endophyte effect sizes on Chl under non-stressed and stressed conditions. Error bars indicate ±95% confidence intervals of the means calculated by the Hedge's method [70]. If the error bars overlap the *dotted line* (effect size = 0.0), the effect is considered nonsignificant. The *light* and *dark grey shades* represent the overall endophyte effect sizes (the aggregates of 72 and 62 studies, corresponding to one in Fig. 4.2) under non-stressed and stressed conditions

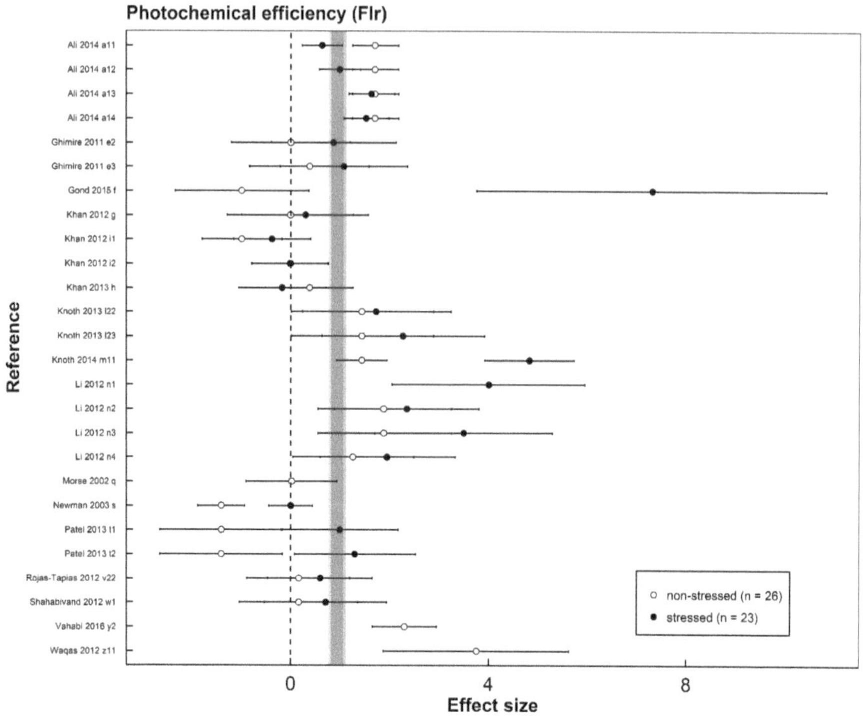

Fig. A2 Effect sizes of various endophytes on host plants' photochemical efficiency (Flr). *Open* and *closed circles* indicate the mean endophyte effect sizes on Flr under non-stressed and stressed conditions. Error bars indicate ±95% confidence intervals of the means calculated by the Hedge's method [70]. If the error bars overlap the *dotted line* (effect size = 0.0), the effect is considered nonsignificant. The *light* and *dark grey shades* represent the overall endophyte effect sizes (the aggregates of 26 and 23 studies, corresponding to one in Fig. 4.2) under non-stressed and stressed conditions

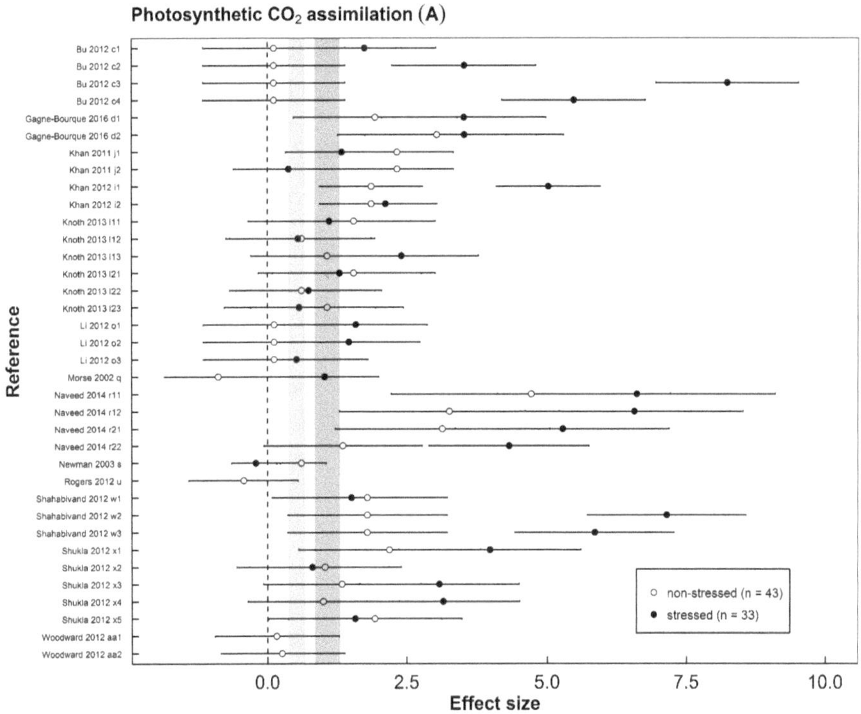

Fig. A3 Effect sizes of various endophytes on host plants' net CO_2 assimilation rate (A). *Open* and *closed circles* indicate the mean endophyte effect sizes on A under non-stressed and stressed conditions. Error bars indicate ±95% confidence intervals of the means calculated by the Hedge's method [70]. If the error bars overlap the *dotted line* (effect size = 0.0), the effect is considered nonsignificant. The *light* and *dark grey shades* represent the overall endophyte effect sizes (the aggregates of 43 and 33 studies, corresponding to one in Fig. 4.2) under non-stressed and stressed conditions

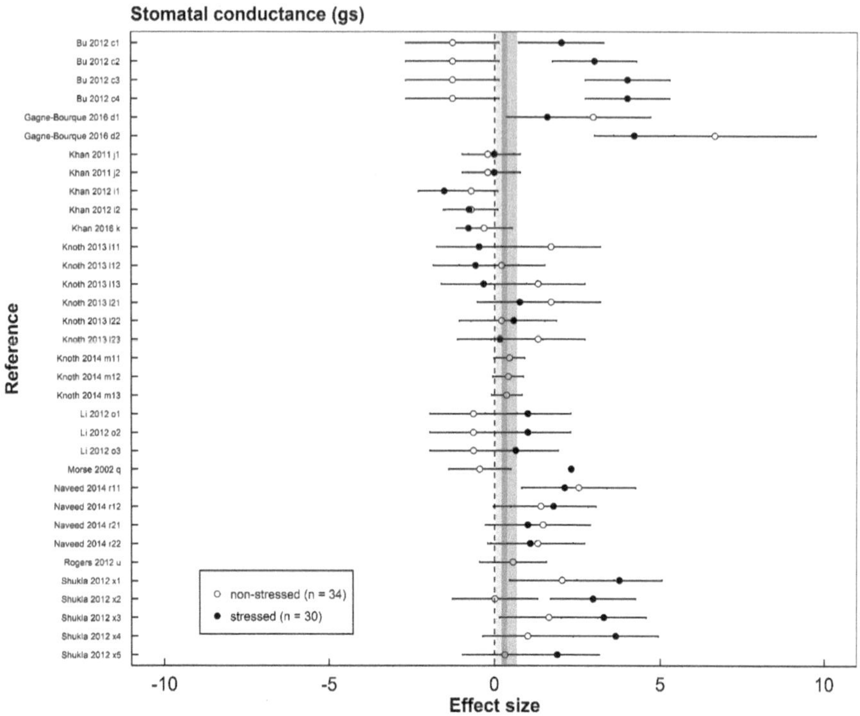

Fig. A4 Effect sizes of various endophytes on host plants' stomatal conductance (gs). *Open* and *closed circles* indicate the mean endophyte effect sizes on gs under non-stressed and stressed conditions. Error bars indicate ±95% confidence intervals of the means calculated by the Hedge's method [70]. If the error bars overlap the *dotted line* (effect size = 0.0), the effect is considered nonsignificant. The *light* and *dark grey shades* represent the overall endophyte effect sizes (the aggregates of 34 and 30 studies, corresponding to one in Fig. 4.2) under non-stressed and stressed conditions

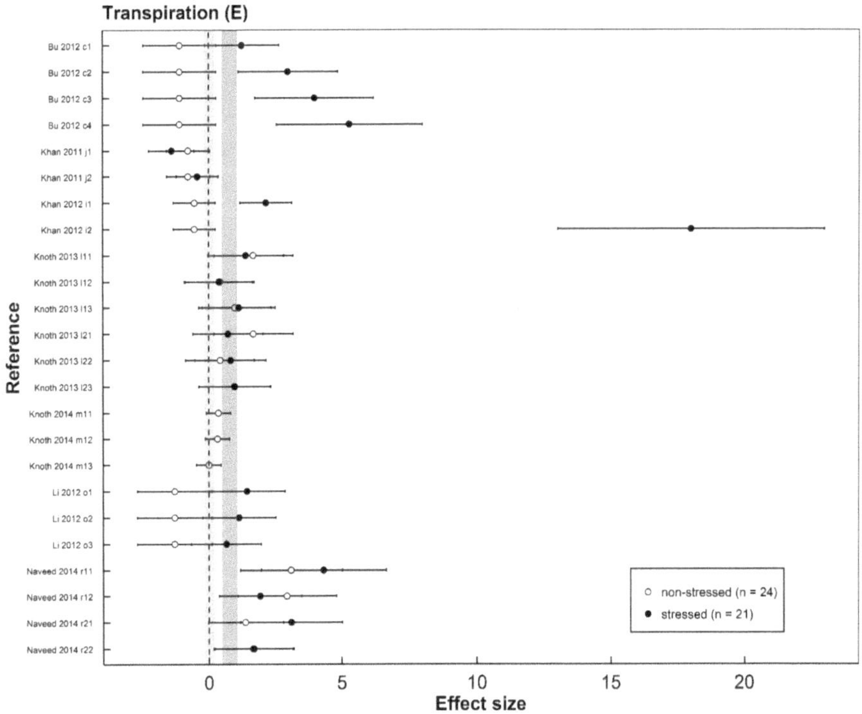

Fig. A5 Effect sizes of various endophytes on host plants' transpiration rate (E). *Open* and *closed circles* indicate the mean endophyte effect sizes on E under non-stressed and stressed conditions. Error bars indicate ±95% confidence intervals of the means calculated by the Hedge's method [70]. If the error bars overlap the *dotted line* (effect size = 0.0), the effect is considered nonsignificant. The *light* and *dark grey shades* represent the overall endophyte effect sizes (the aggregates of 24 and 21 studies, corresponding to one in Fig. 4.2) under non-stressed and stressed conditions

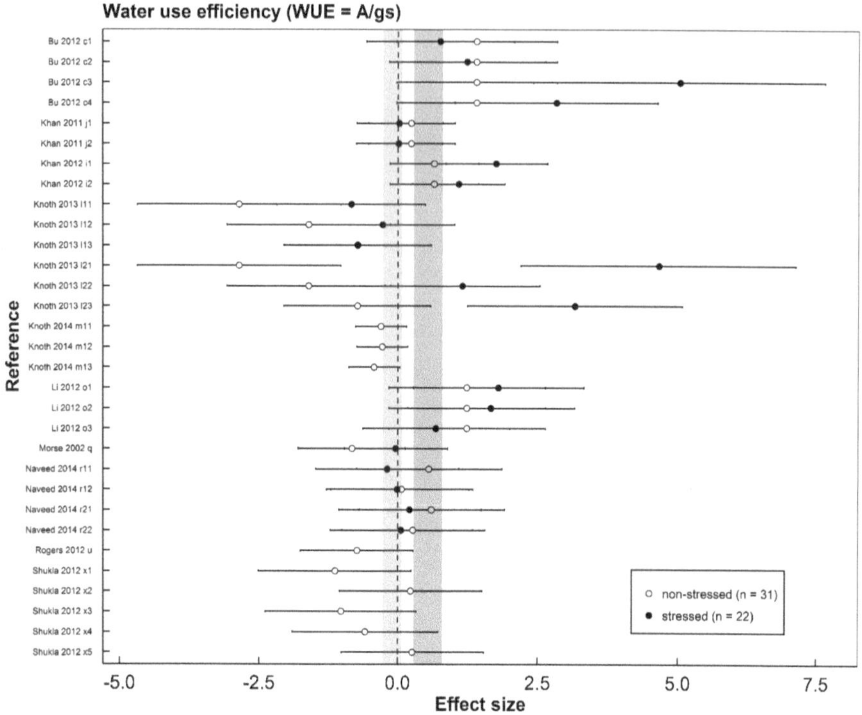

Fig. A6 Effect sizes of various endophytes on host plants' intrinsic water use efficiency (WUE). *Open* and *closed circles* indicate the mean endophyte effect sizes on WUE under non-stressed and stressed conditions. Error bars indicate ±95% confidence intervals of the means calculated by the Hedge's method [70]. If the error bars overlap the *dotted line* (effect size = 0.0), the effect is considered nonsignificant. The *light* and *dark grey shades* represent the overall endophyte effect sizes (the aggregates of 31 and 22 studies, corresponding to one in Fig. 4.2) under non-stressed and stressed conditions

References

1. Dobereiner J. History and new perspectives of diazotrophs in association with nonleguminous plants. Symbiosis. 1992;13:1–13.
2. Doty SL, Sher AW, Fleck ND, Khorasani M, Bumgarner RE, Khan Z, Ko AWK, Kim S-H, DeLuca TH. Variable nitrogen fixation in wild populus. PLoS One. 2016;11:e0155979.
3. Knoth J, Kim S, Ettl G, Doty S. Biological nitrogen fixation and biomass accumulation within poplar clones as a result of inoculations with diazotrophic endophyte consortia. New Phytol. 2014;201:599–609.
4. Barnawal D, Bharti N, Tripathi A, Pandey SS, Chanotiya CS, Kalra A. ACC-deaminase-producing endophyte Brachybacterium paraconglomeratum strain SMR20 ameliorates Chlorophytum salinity stress via altering Phytohormone generation. J Plant Growth Regul. 2016;35:553–64.

5. Waqas M, Khan AL, Kamran M, Hamayun M, Kang SM, Kim YH, Lee IJ. Endophytic fungi produce gibberellins and indoleacetic acid and promotes host-plant growth during stress. Molecules. 2012;17:10754–73.
6. Xin G, Zhang G, Kang J, Staley J, Doty S. A diazotrophic, indole-3-acetic acid-producing endophyte from wild cottonwood. Biol Fertil Soils. 2009;45:669–74.
7. Costa JM, Loper JE. Characterization of siderophore production by the biological control agent Enterobacter cloacae. Mol Plant-Microbe Interact. 1994;7:440.
8. Wakelin SA, Warren RA, Harvey PR, Ryder MH. Phosphate solubilization by Penicillium spp. closely associated with wheat roots. Biol Fertil Soils. 2004;40:36–43.
9. Khan Z, Rho H, Firrincieli A, Hung SH, Luna V, Masciarelli O, Kim S-H, Doty SL. Growth enhancement and drought tolerance of hybrid poplar upon inoculation with endophyte consortia. Curr Plant Biol. 2016;6:38–47.
10. Redman RS, Kim YO, Woodward CJDA, Greer C, Espino L, Doty SL, Rodriguez RJ. Increased fitness of rice plants to abiotic stress via habitat adapted symbiosis: a strategy for mitigating impacts of climate change, PLos One. 2011:6, e14823.
11. Rodriguez RJ, Henson J, Van Volkenburgh E, Hoy M, Wright L, Beckwith F, Kim Y-O, Redman RS. Stress tolerance in plants via habitat-adapted symbiosis. ISME J. 2008;2:404–16.
12. Bulgarelli D, Schlaeppi K, Spaepen S, Ver Loren van Themaat E, Schulze-Lefert P. Structure and functions of the bacterial microbiota of plants. Annu Rev Plant Biol. 2013;64:807–38.
13. Rodriguez R, Freeman D. Symbiotic regulation of plant growth, development and reproduction. Commun Integr Biol. 2009;2:141–3.
14. Ryan RP, Germaine K, Franks A, Ryan DJ, Dowling DN. Bacterial endophytes: recent developments and applications. FEMS Microbiol Lett. 2008;278:1–9.
15. Friesen ML, Porter SS, Stark SC, von Wettberg EJ, Sachs JL, Martinez-Romero E. Microbially mediated plant functional traits. Annu Rev Ecol Evol Syst. 2011;42:23–46.
16. Ali S, Charles TC, Glick BR. Amelioration of high salinity stress damage by plant growth-promoting bacterial endophytes that contain ACC deaminase. Plant Physiol Biochem. 2014;80:160–7.
17. Bu N, Li X, Li Y, Ma C, Ma L, Zhang C. Effects of Na_2CO_3 stress on photosynthesis and antioxidative enzymes in endophyte infected and non-infected rice. Ecotoxicol Environ Saf. 2012;78:35–40.
18. Gagné-Bourque F, Bertrand A, Claessens A, Aliferis KA, Jabaji S. Alleviation of drought stress and metabolic changes in timothy (Phleum pratense L.) colonized with Bacillus subtilis B26. Front Plant Sci. 2016;7:1–16.
19. Ghimire SR, Craven KD. Enhancement of switchgrass (Panicum virgatum L.) biomass production under drought conditions by the ectomycorrhizal fungus Sebacina vermifera. Appl Environ Microbiol. 2011;77:7063–7.
20. Gond SK, Torres MS, Bergen MS, Helsel Z, White JF. Induction of salt tolerance and up-regulation of aquaporin genes in tropical corn by rhizobacterium *Pantoea agglomerans*. Lett Appl Microbiol. 2015;60:392–9.
21. Khan AL, Hamayun M, Khan SA, Kang SM, Shinwari ZK, Kamran M, ur Rehman S, Kim JG, Lee IJ. Pure culture of Metarhizium anisopliae LHL07 reprograms soybean to higher growth and mitigates salt stress. World J Microbiol Biotechnol. 2012;28:1483–94.
22. Khan AL, Lee I-J. Endophytic Penicillium funiculosum LHL06 secretes gibberellin that reprograms Glycine max L. growth during copper stress. BMC Plant Biol. 2013;13:86.
23. Khan Z, Guelich G, Phan H, Redman R, Doty S. Bacterial and yeast endophytes from poplar and willow promote growth in crop plants and grasses. ISRN Agron. 2012;2012:1–11.
24. Khan AL, Hamayun M, Kim Y-H, Kang S-M, Lee I-J. Ameliorative symbiosis of endophyte (Penicillium funiculosum LHL06) under salt stress elevated plant growth of Glycine max L. Plant Physiol Biochem. 2011;49:852–61.
25. Knoth JL, Kim S-H, Ettl GJ, Doty SL. Effects of cross host species inoculation of nitrogen-fixing endophytes on growth and leaf physiology of maize. Glob Change Biol Bioenergy. 2013;5:408–18.

26. Li X, Ren A, Han R, Yin L, Wei M, Gao Y. Endophyte-mediated effects on the growth and physiology of Achnatherum sibiricum are conditional on both N and P availability. PLos One. 2012:7, e48010.
27. Li X, Bu N, Li Y, Ma L, Xin S, Zhang L. Growth, photosynthesis and antioxidant responses of endophyte infected and non-infected rice under lead stress conditions. J Hazard Mater. 2012;213:55–61.
28. Madhaiyan M, Peng N, Te NS, Hsin IC, Lin C, Lin F, Reddy C, Yan H, Ji L. Improvement of plant growth and seed yield in Jatropha curcas by a novel nitrogen-fixing root associated Enterobacter species. Biotechnol Biofuels. 2013;6:140.
29. Morse LJ, Day TA, Faeth SH. Effect of Neotyphodium endophyte infection on growth and leaf gas exchange of Arizona fescue under contrasting water availability regimes. Environ Exp Bot. 2002;48:257–68.
30. Naveed M, Mitter B, Reichenauer TG, Wieczorek K, Sessitsch A. Increased drought stress resilience of maize through endophytic colonization by Burkholderia phytofirmans PsJN and Enterobacter sp. FD17. Environ Exp Bot. 2014;97:30–9.
31. Newman JA, Abner ML, Dado RG, Gibson DJ, Brookings A, Parsons AJ. Effects of elevated CO_2, nitrogen and fungal endophyte-infection on tall fescue: growth, photosynthesis, chemical composition and digestibility. Glob Chang Biol. 2003;9:425–37.
32. Patel D, Saraf M. Influence of soil ameliorants and microflora on induction of antioxidant enzymes and growth promotion of Jatropha curcas L. under saline condition. Eur J Soil Biol. 2013;55:47–54.
33. Rogers A, McDonald K, Muehlbauer MF, Hoffman A, Koenig K, Newman L, Taghavi S, van der Lelie D. Inoculation of hybrid poplar with the endophytic bacterium Enterobacter sp 638 increases biomass but does not impact leaf level physiology. Glob Change Biol Bioenergy. 2012;4:364–70.
34. Rojas-Tapias D, Moreno-Galván A, Pardo-Díaz S, Obando M, Rivera D, Bonilla R. Effect of inoculation with plant growth-promoting bacteria (PGPB) on amelioration of saline stress in maize (Zea mays). Appl Soil Ecol. 2012;61:264–72.
35. Shahabivand S, Maivan HZ, Goltapeh EM, Sharifi M, Aliloo AA. The effects of root endophyte and arbuscular mycorrhizal fungi on growth and cadmium accumulation in wheat under cadmium toxicity. Plant Physiol Biochem. 2012;60:53–8.
36. Shukla N, Awasthi RP, Rawat L, Kumar J. Biochemical and physiological responses of rice (Oryza sativa L.) as influenced by Trichoderma harzianum under drought stress. Plant Physiol Biochem. 2012;54:78–88.
37. Vahabi K, Dorcheh SK, Monajembashi S, Westermann M, Reichelt M, Falkenberg D, Hemmerich P, Sherameti I, Oelmuller R. Stress promotes Arabidopsis-Piriformospora indica interaction. Plant Signal Behav. 2016;11:e1136763.
38. Woodward C, Hansen L, Beckwith F, Redman RS, Rodriguez RJ. Symbiogenics: an epigenetic approach to mitigating impacts of climate change on plants. Hortscience. 2012;47:699–703.
39. Yang B, Wang X-M, Ma H-Y, Li X, Dai C-C. Effects of the fungal endophyte Phomopsis liquidambari on nitrogen uptake and metabolism in rice. Plant Growth Regul. 2014;73:165–79.
40. Zarea MJ, Hajinia S, Karimi N, Mohammadi Goltapeh E, Rejali F, Varma A. Effect of Piriformospora indica and Azospirillum strains from saline or non-saline soil on mitigation of the effects of NaCl. Soil Biol Biochem. 2012;45:139–46.
41. Zhang YP, Nan ZB. Growth and anti-oxidative systems changes in Elymus dahuricus is affected by Neotyphodium endophyte under contrasting water availability. J Agron Crop Sci. 2007;193:377–86.
42. Straub D, Rothballer M, Hartmann A. The genome of the endophytic bacterium H. frisingense GSF30(T) identifies diverse strategies in the Herbaspirillum genus to interact with plants. Front Microbiol. 2013;4:168.
43. Ghabooli M, Khatabi B, Ahmadi FS, Sepehri M, Mirzaei M, Amirkhani A, Jorrín-Novo JV, Salekdeh GH. Proteomics study reveals the molecular mechanisms underlying water stress tolerance induced by Piriformospora indica in barley. J Proteome. 2013;94:289–301.
44. Saikkonen K, Faeth SH, Helander M, Sullivan TJ. Fungal endophytes: a continuum of interactions with host plants. Annu Rev Ecol Syst. 1998;29:319–43.

45. Khan AL, Hamayun M, Waqas M. Exophiala sp. LHL08 association gives heat stress tolerance by avoiding oxidative damage to cucumber plants. Biol Fertil Soils. 2012;48:519–29.
46. Kandel SL, Herschberger N, Kim SH, Doty SL. Diazotrophic endophytes of poplar and willow for growth promotion of rice plants in nitrogen-limited conditions. Crop Sci. 2015;55:1765.
47. Khan Z, Kandel S, Ramos D, Ettl G, Kim S-H, Doty S. Increased biomass of nursery-grown Douglas-fir seedlings upon inoculation with diazotrophic endophytic consortia. Forests. 2015;6:3582–93.
48. Hunt R, Causton DR, Shipley B, Askew AP. A modern tool for classical plant growth analysis. Ann Bot. 2002;90:485–8.
49. Lichtenthaler HK. [34] Chlorophylls and carotenoids: pigments of photosynthetic biomembranes. Methods Enzymol. 1987;148:350–82.
50. Cerovic ZG, Masdoumier G, Ben Ghozlen N, Latouche G. A new optical leaf-clip meter for simultaneous non-destructive assessment of leaf chlorophyll and epidermal flavonoids. Physiol Plant. 2012;146:251–60.
51. Baker NR. Chlorophyll fluorescence: a probe of photosynthesis in vivo. Annu Rev Plant Biol. 2008;59:89–113.
52. Taiz L, Zeiger E. Plant physiology. Sunderland, MA: Sinauer; 2010.
53. Long SP, Bernacchi CJ. Gas exchange measurements, what can they tell us about the underlying limitations to photosynthesis? Procedures and sources of error. J Exp Bot. 2003;54:2393–401.
54. Woodward FI, Lake JA, Quick WP. Stomatal development and CO_2: ecological consequences. New Phytol. 2002;153:477–84.
55. Bunce JA. How do leaf hydraulics limit stomatal conductance at high water vapour pressure deficits? Plant Cell Environ. 2006;29:1644–50.
56. Hetherington AM, Woodward FI. The role of stomata in sensing and driving environmental change. Nature. 2003;424:901–8.
57. Tardieu F, Davies WJ. Stomatal response to abscisic acid is a function of current plant water status. Plant Physiol. 1992;98:540–5.
58. Melotto M, Underwood W, Koczan J, Nomura K, He SY. Plant stomata function in innate immunity against bacterial invasion. Cell. 2006;126:969–80.
59. Milla R, De Diego-Vico N, Martín-Robles N. Shifts in stomatal traits following the domestication of plant species. J Exp Bot. 2013;64:3137–46.
60. Farquhar GD, Sharkey TD. Stomatal conductance and photosynthesis. Annu Rev Plant Physiol. 1982;33:317–45.
61. Damour G, Simonneau T, Cochard H, Urban L. An overview of models of stomatal conductance at the leaf level. Plant Cell Environ. 2010;33:1419–38.
62. Snyder KA, Richards JH, Donovan LA. Night-time conductance in C3 and C4 species: do plants lose water at night? J Exp Bot. 2003;54:861–5.
63. Rosegrant MW, Ringler C, Zhu T. Water for agriculture: maintaining food security under growing scarcity. Annu Rev Environ Resour. 2009;34:205–22.
64. Bernacchi CJ, VanLoocke A. Terrestrial ecosystems in a changing environment: a dominant role for water. Annu Rev Plant Biol. 2015;66:599–622.
65. Lambers H, Chapin FS, Pons TL, Ecology PP, Transport LD, Associations P, Balance E, Plants IA, Organs I, Irradiance E, et al. Plant physiological ecology. New York, NY: Springer New York; 2008.
66. Berg G. Plant-microbe interactions promoting plant growth and health: perspectives for controlled use of microorganisms in agriculture. Appl Microbiol Biotechnol. 2009;84:11–8.
67. Wani ZA, Ashraf N, Mohiuddin T, Riyaz-Ul-Hassan S. Plant-endophyte symbiosis, an ecological perspective. Appl Microbiol Biotechnol. 2015;99:2955–65.
68. Newsham KK. A meta-analysis of plant responses to dark septate root endophytes. New Phytol. 2011;190:783–93.
69. Omacini M, Semmartin M, Pérez LI, Gundel PE. Grass–endophyte symbiosis: a neglected aboveground interaction with multiple belowground consequences. Appl Soil Ecol. 2012;61:273–9.

70. Hedges LV, Gurevitch J, Curtis PS. The meta-analysis of response ratios in experimental ecology. Ecology. 1999;80:1150–6.
71. Anzhi R, Yubao G, Wei W, Jinlong W. Photosynthetic pigments and photosynthetic products of endophyte-infected and endophyte-free Lollum perenne L. under drought stress conditions. Front Biol China. 2006;2:168–73.
72. Rodriguez RJ, White JF, Arnold AE, Redman RS. Fungal endophytes: diversity and functional roles. New Phytol. 2009;182:314–30.
73. Ren A, Clay K. Impact of a horizontally transmitted endophyte, Balansia henningsiana, on growth and drought tolerance of Panicum rigidulum. Int J Plant Sci. 2009;170:599–608.
74. Xu G, Fan X, Miller AJ. Plant nitrogen assimilation and use efficiency. Ann. Rev. Plant Biol. 2012;63:153–82.
75. Xu C, Fisher R, Wullschleger SD, Wilson CJ, Cai M, McDowell NG. Toward a mechanistic modeling of nitrogen limitation on vegetation dynamics. PLoS One. 2012;7:e37914.
76. Yang B, Wang XM, Ma HY, Jia Y, Li X, Dai CC. Effects of the fungal endophyte Phomopsis liquidambari on nitrogen uptake and metabolism in rice. Plant Growth Regul. 2014;73:165–79.
77. Poorter H, Remkes C, Lambers H. Carbon and nitrogen economy of 24 wild-species differing in relative growth-rate. Plant Physiol. 1990;94:621–7.
78. Evans J. Photosynthesis and nitrogen relationships in leaves of C3 plants. Oecologia. 1989;78:9–19.
79. Straub D, Yang H, Liu Y, Tsap T, Ludewig U. Root ethylene signalling is involved in Miscanthus sinensis growth promotion by the bacterial endophyte Herbaspirillum frisingense GSF30 T. J Exp Bot. 2013;64:4603–15.
80. Netto AT, Campostrini E, De Oliveira JG, Bressan-Smith RE. Photosynthetic pigments, nitrogen, chlorophyll a fluorescence and SPAD-502 readings in coffee leaves. Sci Hortic. 2005;104:199–209.
81. Grassi G, Magnani F. Stomatal, mesophyll conductance and biochemical limitations to photosynthesis as affected by drought and leaf ontogeny in ash and oak trees. Plant Cell Environ. 2005;28:834–49.
82. Bae H, Sicher RC, Kim MS, Kim S-H, Strem MD, Melnick RL, Bailey B a. The beneficial endophyte Trichoderma hamatum isolate DIS 219b promotes growth and delays the onset of the drought response in Theobroma cacao. J Exp Bot. 2009;60:3279–95.
83. Parry MAJ, Andralojc PJ, Mitchell RAC, Madgwick PJ, Keys AJ. Manipulation of Rubisco: the amount, activity, function and regulation. J Exp Bot. 2003;54:1321–33.
84. Costa Pinto LSR, Azevedo JL, Pereira JO, Carneiro Vieira ML, Labate CA. Symptomless infection of banana and maize by endophytic fungi impairs photosynthetic efficiency. New Phytol. 2000;147:609–15.
85. Belesky DP, Devine OJ, Pallas JE Jr, Stringer WC. Photosynthetic activity of tall fescue as influenced by a fungal endophyte. Photosynthetica. 1987;21:82–7.
86. Evans J, Terashima I. Effects of nitrogen nutrition on electron transport components and photosynthesis in spinach. Funct Plant Biol. 1987:59–68.
87. Herridge DF, Peoples MB, Boddey RM. Global inputs of biological nitrogen fixation in agricultural systems. Plant Soil. 2008;311:1–18.
88. Evans JR. Improving photosynthesis. Plant Physiol. 2013;162:1780–93.
89. Ort DR, Merchant SS, Alric J, Barkan A, Blankenship RE, Bock R, Croce R, Hanson MR, Hibberd JM, Long SP, et al. Redesigning photosynthesis to sustainably meet global food and bioenergy demand. Proc Natl Acad Sci U S A. 2015;112:1–8.
90. Xin C-P, Tholen D, Devloo V, Zhu X-G. The benefits of photorespiratory bypasses: how can they work? Plant Physiol. 2015;167:574–85.
91. Zhu X-G, Long SP, Ort DR. Improving photosynthetic efficiency for greater yield. Annu Rev Plant Biol. 2010;61:235–61.
92. Rogers A, Ainsworth EA, Leakey ADB. Will elevated carbon dioxide concentration amplify the benefits of nitrogen fixation in legumes? Plant Physiol. 2009;151:1009–16.
93. Lemoine R, La Camera S, Atanassova R, Dédaldéchamp F, Allario T, Pourtau N, Bonnemain J-L, Laloi M, Coutos-Thévenot P, Maurousset L, et al. Source-to-sink transport of sugar and regulation by environmental factors. Front Plant Sci. 2013;4:272.

94. Jones DL, Nguyen C, Finlay RD. Carbon flow in the rhizosphere: carbon trading at the soil-root interface. Plant Soil. 2009;321:5–33.
95. Lambers H, Atkin O, Millenaar F. Respiratory patterns in roots in relation to their functioning. In: Plant roots. Boca Raton, FL: CRC Press; 2002. p. 521–52.
96. Baas R, van der Werf A, Lambers H. Root respiration and growth in Plantago major as affected by vesicular-arbuscular mycorrhizal infection. Plant Physiol. 1989;91:227–32.
97. Kaschuk G, Hungria M, Leffelaar PA, Giller KE, Kuyper TW. Differences in photosynthetic behaviour and leaf senescence of soybean (Glycine max [L.] Merrill) dependent on N2 fixation or nitrate supply. Plant Biol. 2010;12:60–9.
98. Kiers ET, Duhamel M, Beesetty Y, Mensah JA, Franken O, Verbruggen E, Fellbaum CR, Kowalchuk GA, Hart MM, Bago A, et al. Reciprocal rewards stabilize cooperation in the mycorrhizal symbiosis. Science (New York, NY). 2011;333:880–2.
99. Kiers ET, Rousseau RA, West SA, Denison RF. Host sanctions and the legume-rhizobium mutualism. Nature. 2003;425:1095–8.
100. Malinowski DP, Belesky DP. Adaptations of endophyte-infected cool-season grasses to environmental stresses: mechanisms of drought and mineral stress tolerance. Crop Sci. 2000;40:923–40.
101. Barker DJ, Hume DE, Quigley PE. Negligible physiological responses to water deficit in endophyte-infected and uninfected perennial ryegrass. In: Bacon CW, Hill NS, editors. Neotyphodium/grass interactions. New York and London: Plenum Press; 1997. p. 137–9.
102. Buck GW, West CP, Elbersen HW. Endophyte effect on drought tolerance in diverse Festuca species. In: Bacon CW, Hill NS, editors. Neotyphodium/grass interactions. New York and London: Plenum Press; 1997. p. 141–3.
103. Elmi AA, West CP. Endophyte infection effects on stomatal conductance, osmotic adjustment and drought recovery of tall fescue. New Phytol. 1995;131:61–7.
104. Richardson MD, Hoveland CS, Bacon CW. Photosynthesis and stomatal conductance of symbiotic and nonsymbiotic tall fescue. Crop Sci. 1993;33:145–9.
105. Turner N. Adaptation to water deficits: a changing perspective. Aust J Plant Physiol. 1986;13:175.
106. Tiwari R, Awasthi A, Mall M, Shukla AK, Srinivas KVNS, Syamasundar KV, Kalra A. Bacterial endophyte-mediated enhancement of in planta content of key terpenoid indole alkaloids and growth parameters of Catharanthus roseus. Ind Crop Prod. 2013;43:306–10.
107. Augé RM, Toler HD, Saxton AM. Arbuscular mycorrhizal symbiosis alters stomatal conductance of host plants more under drought than under amply watered conditions: a meta-analysis. Mycorrhiza. 2014;25:13–24.
108. Worchel ER, Giauque HE, Kivlin SN. Fungal symbionts alter plant drought response. Microb Ecol. 2013;65:671–8.
109. Franks PJ, W. Doheny-Adams T, Britton-Harper ZJ, Gray JE. Increasing water-use efficiency directly through genetic manipulation of stomatal density. New Phytol. 2015;207:188–95.
110. Nassar AH, El-Tarabily KA, Sivasithamparam K. Promotion of plant growth by an auxin-producing isolate of the yeast Williopsis saturnus endophytic in maize (Zea mays L.) roots. Biol Fertil Soils. 2005;42:97–108.
111. Xin G, Zhang G, Kang JW, Staley JT, Doty SL. A diazotrophic, indole-3-acetic acid-producing endophyte from wild cottonwood. Biol Fertil Soils. 2009;45:669–74.
112. Davitt AJ, Chen C, Rudgers JA. Understanding context-dependency in plant-microbe symbiosis: the influence of abiotic and biotic contexts on host fitness and the rate of symbiont transmission. Environ Exp Bot. 2011;71:137–45.
113. Eerens JPJ, Lucas RJ, Easton S, White JGH. Influence of the endophyte ( ) on morphology, physiology, and alkaloid synthesis of perennial ryegrass during high temperature and water stress. N Z J Agric Res. 2010;41(2):219–26.

Chapter 5
The Symbiogenic Tango: Achieving Climate-Resilient Crops Via Mutualistic Plant-Fungal Relationships

Regina S. Redman and Rusty J. Rodriguez

5.1 Introduction

Since plants lack locomotion, they must adapt to a variety of habitat-imposed stresses such as low nutrients, temperature extremes, high salt levels [1], and water limitations and excess [1–7]. Over the last several decades the frequency and severity of temperature and water stress have increased due to climate change (www.copenhagendiagnosis.com) and there have been no commercially available technologies to adequately address these issues, until recently (www.adsymtech.com).

The ability of plants to adapt to abiotic stress is thought to involve genetic processes limited to the plant genome [6, 8–10]. Yet, the mechanisms responsible for stress tolerance are poorly defined. For example, all plants sense and respond to temperature, water, and chemical stress, but few species are able to colonize high-stress habitats [5, 11–16]. Although there are many reports on how plants respond to stress, the underlying mechanisms that allow plants to thrive in high-stress habitats remain unresolved [1–3, 5, 6, 12, 13, 15, 17–25].

Current views on plant adaptation to stress assume that plants exist and operate as individuals. However, all plants in nature are symbiotic with fungi and other microorganisms. Over the last 20 years it has become clear that at least one group of plant symbiotic fungi, known as fungal endophytes (defined below), is responsible for the adaptation and survival of plants in high-stress habitats [5, 13, 14, 19, 26, 27].

R.S. Redman (✉)
Adaptive Symbiotic Technologies, Seattle, WA 98105, USA

Symbiogenics, Seattle, WA 98105, USA
e-mail: rsredman@adsymtech.com

R.J. Rodriguez
Adaptive Symbiotic Technologies, Seattle, WA 98105, USA

Symbiogenics, Seattle, WA 98105, USA

Department of Biology, University of Washington, Seattle, WA 98195-5325, USA

S.L. Doty (ed.), *Functional Importance of the Plant Microbiome*,
DOI 10.1007/978-3-319-65897-1_5

This paradigm changes how plants adapt to stress and opens new opportunities to mitigate impacts of climate change in natural and agricultural ecosystems.

An array of symbiotic lifestyles may be expressed by plant-microbe associations and the outcome is defined based on fitness benefits (positive, neutral, or negative) experienced by one or both partners [26, 28, 29]. This chapter focuses on mutualistic symbioses, where both partners in the relationship experience positive fitness benefits, and how these associations can be used to enhance agricultural productivity.

5.2 Symbiotic Fungal Endophytes

Fungal endophytes are defined as living within ("endo") the plant ("phyte") and are divided into four main classes dependent upon their particular-spatial orientation within the plants, horizontal or vertical transmission mode, benefits conferred, and plant host range [27].

Of all the fungal endophytes, Class 1 endophytes have been the most well studied. They are comprised of a limited number of fungal genera, able to colonize all parts of a plant, are vertically and horizontally transmitted, difficult to propagate axenically, and have a limited host range consisting primarily of grass species [27, 30–35]. In contrast, Class 2 endophytes are comprised of a large number of fungal genera with a broad host range, reside within the vegetative tissues, are vertically and horizontally transmitted, and are easy to propagate axenically. Although there has been a growing body of knowledge in recent years, this group is still largely understudied [4, 5, 23, 24, 27, 28, 36–40].

Class 3 endophytes are comprised of a hyper-diverse number of fungal species, are distinguished based on their highly localized occurrence in above-ground tissues such as the leaves, have a large host range with transmission occurring horizontally. Transmission can also occur via wind or rain splash dispersal. Class 3 endophytes on tropical plant leaves appear to play an important ecological role providing both biotic and abiotic stress tolerance. However, as yet, this system has caught the attention of a limited number of researchers and merits more study [27, 41–43].

Last but not the least are the mysterious Class 4 endophytes. These endophytes are abundantly associated solely with the roots of numerous herbaceous and woody plant species, over a vast geographical range. First observed by Merlin in 1922, this class of fungal endophytes are referred to as dark septate endophytes (DSE) due to the high level of pigmentation observed in the fungal mycelium. Because of the abundance of DSE in a wide diversity of plant species, it is inferred that this group must play an important role in plant adaptation and survival [27, 44–46]. Indeed, studies conducted by researchers in India with DSE isolated from native desert plants (subjected to elevated temperature and drought stress) demonstrated that the same stress benefits could be imparted to medicinal plants symbiotically [47]. This group requires additional attention as the extent of the functional role(s), identification of the partners in the association, and mode of transmission are largely uncharacterized and remain a mystery [27, 45].

5.3 Endophytes and Stress Tolerance

Class 1 endophytes are unique in that they represent a small number of phylogenetically related clavicipitaceous fungal species (e.g., *Neotyphodium* spp.—teleomorph *Epichloë* spp., *Balansia* spp., *Claviceps* spp.) with a limited host range encompassing some cool and warm season grasses. First noted in Europe in the late nineteenth century in grass plants (*Lolium* spp.,), investigators linked the consumption of endophyte-infected (*Neotyphodium* spp.; teleomorph *Epichloë*) grass with the occurrence of fescue toxicosis in animals such as cattle [5]. This created a debilitating situation for animal husbandry with symptoms including weight loss, increased susceptibility to heat stress, reduced calving rates, low average daily gain (ADG), decreased milk production, and possible death due to ingestion of various fungal derived alkaloid toxic compounds (http://extension.missouri.edu/p/G4669) [48–53]. During the 1800s, tall fescue and their toxin-producing endophytes were unintentionally introduced to the United States. Tall fescue gained popularity and was planted throughout the United States during the 1940s and 1950s primarily due to the robust nature that symbiotic plants presented in the face of various abiotic stress tolerances such as temperature, drought, and salinity. In addition, biotic stress tolerance in the form of anti-herbivory against insects and nematodes and microbial pathogen disease resistance were traits observed with this symbiotic grass [27, 30–34, 54–57].

With such an array of benefits expressed by tall fescue and the ability of the fungal endophyte to be horizontally (via vegetative tissues) and vertically (through the seed embryo) transmitted, it is not surprising that an estimated >90% of pastures in the United States are infected with Class 1, alkaloid toxin-producing, fungal endophytes [27, 34, 35].

Strategies for utilizing such symbiotic systems for the benefit of agriculture are clear, the ultimate goal being conferring all the positive stress benefits to forage and other agricultural systems, without the negative effects of the toxins. To begin addressing these problems, the primary approach was a simple one: Replace endophyte (E+)-colonized seeds with endophyte-free grass seeds (E−). Unfortunately, endophyte removal resulted in reduced seedling vigor and lower plant persistence. As such, much effort has been invested to study the complex biochemical processes and genetic basis of endophyte toxin production [27, 58–61].

The challenges for development of Class 1 endophytes for agricultural application are confounded not only due to toxin production, but also manipulation and axenic growth of these endophytes can be somewhat slow, and laborious. In addition, the host range and application of this group of endophytes are primarily limited to grasses. However, much work has been undertaken to address these issues as the benefits arising from such studies would enhance the forage crop and hence animal husbandry industries.

Taking a more ecological and real-time approach, New Zealand researchers found naturally occurring endophytes that produced alkaloids against insect persistence, but did not produce alkaloids impacting poor animal performance.

These "novel," "animal-friendly" endophytes present great promise as they directly address the concerns of the farmer, in a timely manner. Commercial release of these novel endophytes (MaxQ) is showing great promise as the benefits to the farmer are three-pronged: MaxQ grass plants provide plant vigor, and enhanced yields compared to E− plants; do not negatively impact forage animals; and provide protection against insect herbivory. With all these positive attributes, this Class 1 symbiotic system represents the first commercially produced viable option for combating the negative effects of fescue toxicosis. In addition, MaxQ imparts other positive abiotic and biotic benefits to grasses such as drought resistance and disease protection, respectively. A broader application of this technology to other agricultural monocots such as cereals and maize (*Zea mays* L.) has been demonstrated to be possible and currently under development (www.pennington.com/all-products/agriculture-wildlife/forage/maxq).

Class 2 fungal endophytes are non-clavicipitaceous, and represent a broad range of fungal genera, with a broad host range encompassing both monocot and eudicot plants. These fungi colonize all plant vegetative tissue including the roots, shoots, and seed coats, but not the embryo. As such, this group of endophytes is transmitted horizontally, but can also be transmitted vertically as the maternal plant's seed coat contains the endophyte, and upon seed germination is colonized by the endophyte [27]. One of the more appealing features to this class of endophytes is that nonsymbiotic plants can be easily generated via surface sterilization of seeds. Culturable Class 2 endophytes can be propagated on a variety of simple media, induced to produce asexual single-celled spores (conidia) in abundance, and maintained as long-term stocks in a cost-effective manner. In addition, these fungi can colonize a wide array of plants simply by exposing the seeds or seedlings to the endophyte of interest. Collectively, these traits make for an ideal "model system" for scientific study as both symbiotic and nonsymbiotic plants may be easily generated for testing "Koch's" postulates to identify symbiotic functionality [4, 23, 39, 40].

The first description of a Class 2 endophyte symbiosis occurred more than 100 years ago with a *Phoma* spp. symbiotic with heather plants (*Calluna vulgaris*). This fungus was found capable of colonizing all parts of the plant including the seed coat. In recent years, *Phoma* spp. have been found to be common root endophytes that confer symbiotic benefits to plants [5, 37, 62–66]. Since the 1970s, more than 1000 papers have been published on endophytes. Initially, much of the studies centered around the classification, distribution, and abundance of endophytes from asymptomatic plant tissues [67, 68]. However, in the past several decades, the focus with Class 2 endophytes has been centered around ecological roles with the intent to utilize this symbiotic technology (symbiogenics) to generate climate-resilient plants for habitat restoration, climate adaptability, and agricultural sustainability [23, 40].

5.4 Habitat-Adapted Symbiosis

Over the last several decades, our research has led to new concepts pertaining to plant adaptation and fitness. The ability of fungal endophytes to confer habitat-specific stress tolerance to plants is a phenomenon we designate habitat-adapted symbiosis [13].

5.4.1 Temperature Stress

The discovery of habitat-adapted symbiosis began in the mid-1990s while studying the fungal community structure in hot geothermal soils of Yellowstone National Park (YNP). In addition to the many different species of fungi found living in these soils, there were also 8–10 plant species thriving there [69]. Most of the plant species grew in areas where the soil temperatures were <50 °C. However, in the very hottest soils (≥50 °C), one dominant plant species thrived—*Dichanthelium lanuginosum,* commonly known as tropical panic grass. Although all plants in native habitats were thought to be symbiotic with Class 2 endophytes, plants in high-stress habitats had not been well studied. Therefore, we decided to analyze these plants for symbiotic associations with fugal endophyte(s) and, if present, determine what ecological role the endophytes played. All of the plants analyzed were colonized with the same Class 2 fungal endophyte (*Curvularia protubera* Cp4666D). Furthermore, when the symbiotic partners were grown separately, neither the fungus nor the plant could survive temperatures above 38 °C. However, symbiotically these partners could survive elevated temperature stress (≥65 °C) [4, 70]. This was the first demonstration that symbiosis, and not the plant alone, adapted plants to stress.

5.4.2 Salt Stress

The Yellowstone study was expanded to plants growing in coastal habitats of the San Juan Islands (SJI) in Washington state where American Dunegrass (*Leymus mollis*) was thriving in beach cobble and the root zones exposed to seawater during high tides. More than 100 plants analyzed from several beach locations were found to be symbiotic with the same Class 2 fungal endophyte (*Fusarium culmorum* FcRed1). Studies were conducted in a similar manner to the YNP study, to determine if the *Fusarium* endophyte was responsible for salt adaptation of Dunegrass plants. Both laboratory and field studies showed that these native plants tolerated up to seawater levels of salinity (≥500 mM NaCl), only when symbiotic with the endophyte. Without its symbiotic partner, the Dunegrass plants were no more adapted to high salt than agricultural crops [5, 13, 23].

Further studies revealed that the YNP fungal endophyte conferred temperature tolerance to plants but not salt tolerance and the coastal fungal endophyte conferred salt but not temperature tolerance. The ability of these fungal endophytes to adapt plants in a stress-specific manner is the basis of habitat-adapted symbiosis (HAS). More importantly, moving the fungi into endophyte-free plants resulted in stress adaptation of plants in 24–36 h. Now that is a very rapid adaptive strategy!

5.4.3 Drought Tolerance and Water Use Efficiency

One of the observations made in the geothermal and coastal habitats was that both habitats imposed a high level of drought stress, especially in the hot summer months. Interestingly, the fungal endophytes from both habitats conferred drought tolerance to both native and agricultural plants. Moreover, plants that are symbiotic with these fungi appear to have higher metabolic efficiency and consume less water (≥50% less depending upon the plant system) [23].

5.4.4 Micro-Habitat-Specific Symbiotic Phenomena

Adjacent to one YNP geothermal site, wild strawberry plants (*Fragaria* spp.) were found growing in non-thermal alpine soils (16–25 °C). Symbiotic analysis of the strawberry plants showed that they contained the same fungal species (*Curvularia protuberata*) found in the geothermal plants. However, laboratory and field testing revealed that the strawberry plant endophyte (*Curvularia protuberate* Cp4667S) did not impart temperature stress tolerance [13, 26, 69]. Similarly, in the SJI costal beach habitat site, the same species of plant (*Leymus mollis*) was found to grow in low-salt soils. Laboratory and field testing revealed that endophytes from plants in the high-salinity soils imparted high salt tolerance while endophytes from Dunegrass thriving in low salt-stress soils conferred low levels of salt stress tolerance. To further assess what appeared to be a habitat by isolate-specific phenomenon, isolates of *Curvularia protuberata* (CpMH206) obtained from non-geothermal plants and *Fusarium culmorum* (Fc18) obtained from non-salt stress plants were purchased from the American Type Culture Collection (ATCC) (www.atcc.org.en/Products/Cells_and_*Microorganisms/Fungi*and_Yeast.aspx) and tested in planta. Neither of the ATCC isolates conferred stress tolerance to plants indicating that this is a habitat x isolate phenomenon [5, 13, 23, 26, 69]. These relationships appear to be a dynamic interplay between plant host and fugal partner that can result in stress tolerance, optimal fitness, and plant survival in high-stress habitats.

5.4.5 Symbiotic Communication Is Ancient

The ability of these endophytes to adapt both grasses (monocots) and broadleaf (eudicots) plants suggests that the symbiotic communication involved in stress adaptation predates the divergence of these plant lineages more than ≥200 MYA [71]. Since fossil records indicate that plant-fungal symbiosis is quite ancient (≥400 MYA), fungi are theorized to have been involved in the movement of plants from an aquatic arena onto land by virtue of conferring stress tolerance [72–75]. Perhaps symbiotically conferred drought tolerance is a residual ancient trait that all endophytes share. Although the level of drought tolerance can vary depending upon the plant host and fungal endophyte, ongoing studies indicate that symbiotic benefits such as drought tolerance and water use efficiency are common symbiotic themes [4, 5, 13, 23, 26, 27, 38, 39, 70].

5.4.6 Development of Symbiotic Technology for Agriculture

Discovering that native plant species were adapting to abiotic stress via symbiotic associations was impetus to determine if symbiotic technology (symbiogenics) could be developed to mitigate climate impacts in agriculture. Initial studies revealed that the YNP fungal endophyte could colonize genetically distant plant species such as watermelons and tomatoes. Remarkably, once the YNP endophyte established a symbiosis with these plants, they became adapted to high temperature stress [70]. Similarly, the coastal endophyte adapted tomatoes and rice to salt stress [23]. The extensive host range of these fungal endophytes and their ability to adapt genetically distant plants to stress opened the door to developing microbial products to generate climate-resilient crops. Based on the original YNP and SJI studies, plants in habitats with different types of chemical and physical stress were analyzed and a large library of symbiotic fungal endophytes amassed. The fungal endophytes belong to taxonomically diverse groups including species of *Colletotrichum*, *Fusarium*, *Curvularia*, *Phoma*, *Alternaria*, *Aspergillus*, *Penicillium*, *Acremonium*, *Trichoderma*, *Drechslera*, and *Montagnulacea*. Studies conducted with fungi within the library allowed for the development of numerous criteria to determine if a strain was conducive for commercial development.

Based on 25 years of research, Adaptive Symbiotic Technologies (AST) has developed the product line *BioEnsure*® to generate climate-resilient crop plants. *BioEnsure*® was developed as liquid formulations containing fungal endophytes that are sprayed onto seeds using the same-seed treatment technology used for applying chemicals on seeds. Once on the seeds, the fungi are dormant until seed germination. After seeds germinate, the fungi activate and establish a symbiosis in 24–36 h.

5.5 Application of Symbiotic Technology in Agriculture

Since 2012, AST has been developing fungal endophytes to generate climate-resilient crop plants. Field testing also began in 2012 to determine the efficacy of *BioEnsure*® in mitigating climate stress on crop production. Fortuitously, it was a perfect year to begin as one of the most severe droughts on record occurred in the Midwestern United States. AST established small plot trials to compare corn production with and without the *BioEnsure*® treatment. The results were very positive with the *BioEnsure*®-treated plants producing up to 85% more yield than untreated plants (Fig. 5.1). The success of 2012 led to an expansion of field testing and formula optimization over the next several years culminating with >1000 field tests in 2017 over a geographic range encompassing 12 countries; multiple ecozones (temperate, subtropical, tropical, and desert); >20 crop species (corn, soy, wheat, barley, cotton, rice, sorghum, peas, dry beans, lentil, tomato, watermelon, millet, mung bean, sesame, guar, okra, peppers, potato, sugar beets); multiple plant hybrids and varieties; and numerous soil types and climate zones. All the field testing was done by independent parties including farmers, contract research organizations (CROs), seed companies, and university faculty. Field studies were designed such that side by side, direct comparisons between *BioEnsure*®-treated and untreated controls could be measured. In addition, seeds were coated with commercial Plant Protection Packages (PPP) commonly used in each geographic location. After 5 years of testing *BioEnsure*®, yield data revealed that there is a crop benefit using *BioEnsure*® that is directly proportional to the level of stress plants experience during the growing season (i.e., as stress increases, so too does the yield benefit). When the stress levels are low, the yield benefit averages are 3–5% but as stress levels increase, yield averages

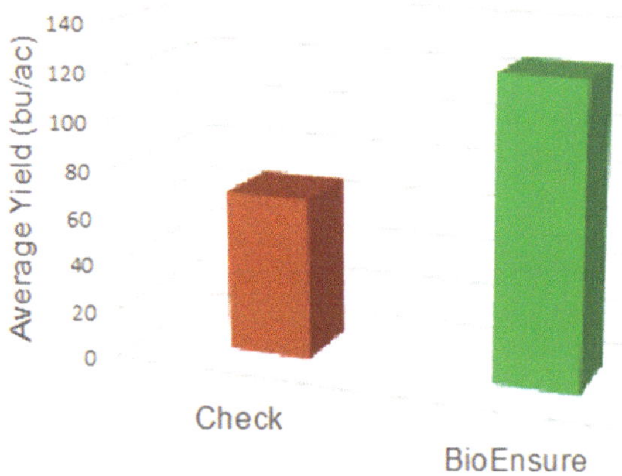

Fig. 5.1 2012 Corn Field Trial Results with Stress. Studies were conducted with hybrid corn and *BioEnsure*® in the presence of elevated temperature and drought stress. A total of two evaluations representing nine replications, across a single location conducted by a farmer. The check/control plants are represented in *red* and the *BioEnsure*®-treated plants in *green*. An average yield increase of 85% (58 bu./ac; 3641 kg/ha) was observed. Seeds were treated with standard PPP containing fungicides, insecticides, polymers, and dyes

increase up to 85%. The potential benefits for commercial agriculture (in the United States as well as abroad) are significant and this technology can be viewed as a yield enhancer that carries an insurance policy against high-stress growing years (Figs. 5.1, 5.2, 5.3, 5.4, 5.5, 5.6, 5.7 and 5.8).

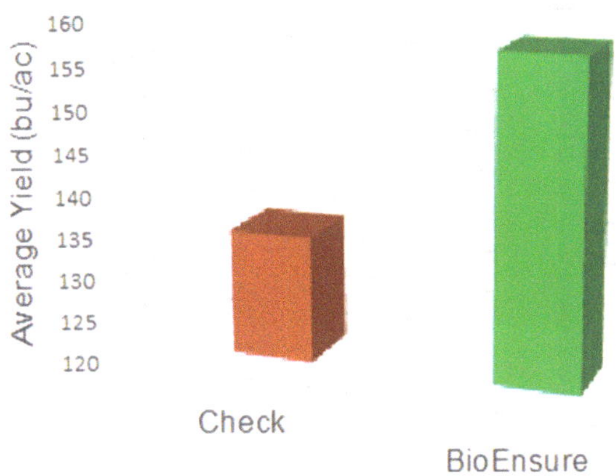

Fig. 5.2 2012–2015 Corn Field Trial Results with Stress. Studies were conducted with *BioEnsure*® in the presence of elevated temperature and drought stress on different hybrid corns ($N = 34$). A total of 15 evaluations representing 45 replications, across nine locations encompassing seven states, were conducted by independent groups such as universities, farmers, and CROs. The check/control plants are represented in *red* and the *BioEnsure*®-treated plants in *green*. An average yield increase of 26% (38.9 bu./ac; 2475 kg/ha) was observed in all evaluations. Of the 15 evaluations, 100% showed yield increases of ≥1% compared to the check/controls. All hybrid seeds tested were treated with standard PPP, the composition and concentration of which could vary depending upon the seed company and could ± include fungicides, insecticides, polymers, and dyes

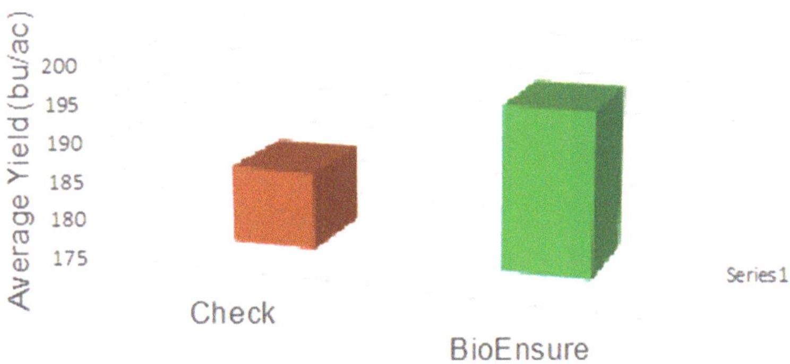

Fig. 5.3 2013–2016 Corn Field Trial Results Without Stress. Studies were conducted with *BioEnsure*® on different corn hybrids ($N = 34$) in the absence of stress. A total of 236 evaluations representing 944 replications, across 62 locations encompassing 16 states, were conducted collectively, by independent groups such as universities, farmers, seed companies, distributors, and CROs. The check/control plants are represented in *red* and the *BioEnsure*®-treated plants in *green*. An average yield increase of 3.5% (6.4 bu./ac; 407 kg/ha) was observed. Of the 236 evaluations, 69% showed yield increases of ≥1% compared to the check/controls. All hybrid seeds tested were treated with standard PPP, the composition and concentration of which could vary depending upon the seed company and could ± include fungicides, insecticides, polymers, and dyes

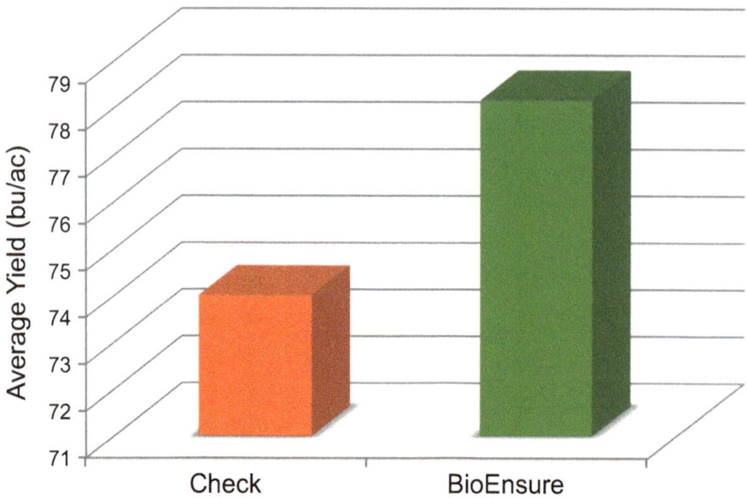

Fig. 5.4 2015–2016 Wheat Field Trial Results Without Stress. Studies were conducted with *BioEnsure*® on five different winter and spring wheat varieties. A total of 22 evaluations representing 66 replications, across 22 locations in seven US states, and one location in Australia, were conducted by independent groups such as universities, farmers, distributors, and CROs. The check/control plants are represented in *red* and the *BioEnsure*®-treated plants in *green*. An average yield increase of 5.6% (4.3 bu./ac; 225 kg/ha) was observed. Of the 22 evaluations, 81.8% showed yield increases of ≥1% compared to the check/controls. All seed varieties tested were treated with standard PPP, the composition and concentration of which could vary depending upon the seed company and could ± include fungicides, insecticides, polymers, and dyes

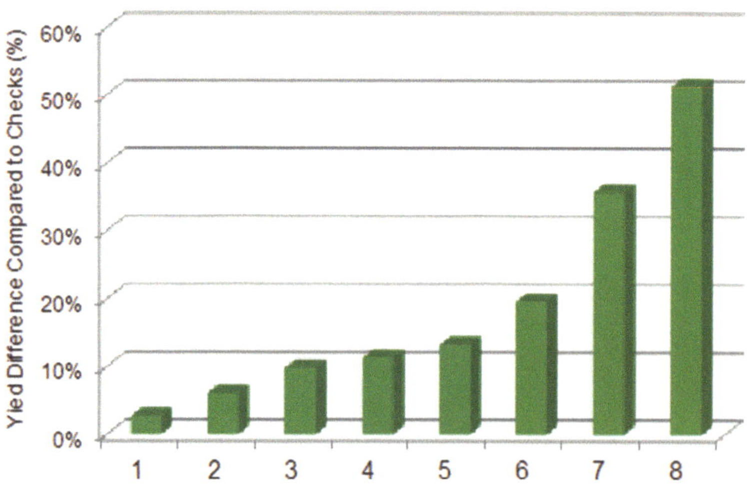

Fig. 5.5 2012–2015 Cotton Field Trial Results with Stress. Studies were conducted with *BioEnsure*® on three cotton varieties in the presence of high temperature and drought stress. A total of eight evaluations representing eight replications in Texas, India, and Australia conducted by farmers and CROs. An average yield increase of 26% was observed with 100% evaluations showing yield increases of ≥1% compared to the check/controls. All seeds tested were treated with standard PPP, the composition and concentration of which could vary depending upon the seed company and could ± include fungicides, insecticides, polymers, and dyes

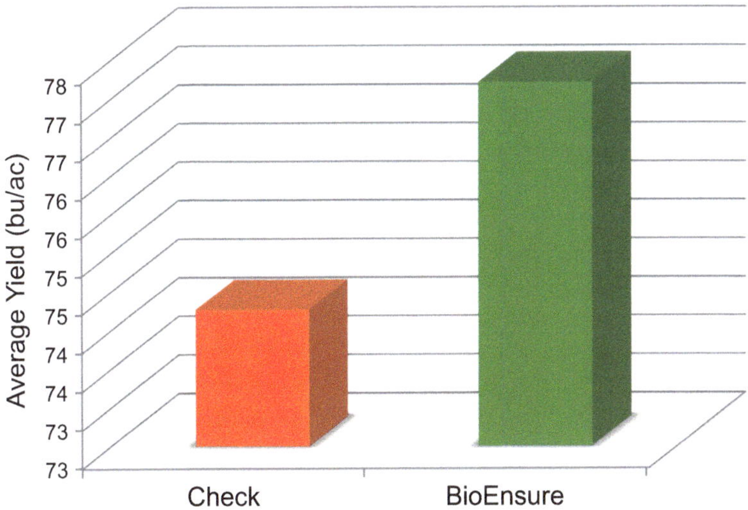

Fig. 5.6 2015–2016 Soybean Field Trial Results Without Stress. Studies were conducted with *BioEnsure®* on five different soybean varieties. A total of 28 evaluations representing 112 replications, across 14 locations in nine US states, and one location in Argentina, were conducted collectively, by independent groups such as universities, seed companies, and CROs. The check/control plants are represented in *red* and the *BioEnsure®*-treated plants in *green*. An average yield increase of 4.3% (3.4 bu./ac; 231 kg/ha) was observed. Of the 28 evaluations, 67.8% showed yield increases of ≥1% compared to the check/controls. All seed varieties tested were treated with standard PPP, the composition and concentration of which could vary depending upon the seed company and could ± include fungicides, insecticides, polymers, and dyes

The greatest benefits of symbiogenic technology are realized in dryland cultivation, especially in regions where temperatures exceed 95 °F (35 °C) during the growing season. For example, field tests on small farms (<2 hectares) in Rajasthan, India, with pearl millet and mung bean resulted in average *BioEnsure®* yield increases of 29% and 55%, respectively (Figs. 5.5 and 5.6). These yield increases are unprecedented and the significance goes beyond crop production. Small landholding farmers grow crops for sustenance, animal fodder, carryover seed for future planting, and revenues. Unless the farms are larger than two hectares there is rarely sufficient yield to generate revenues. The yield benefit of symbiogenic technology on a two-hectare farm equates directly to the % increase in revenues for the farmer. Therefore, this technology can reduce the amount of land required for break-even agricultural production in poor communities. We are hopeful that by combining symbiogenic technology like *BioEnsure®* with other novel technologies, it will be possible for poor farmers to generate significantly more revenues and begin to break the chain of poverty in rural communities around the world.

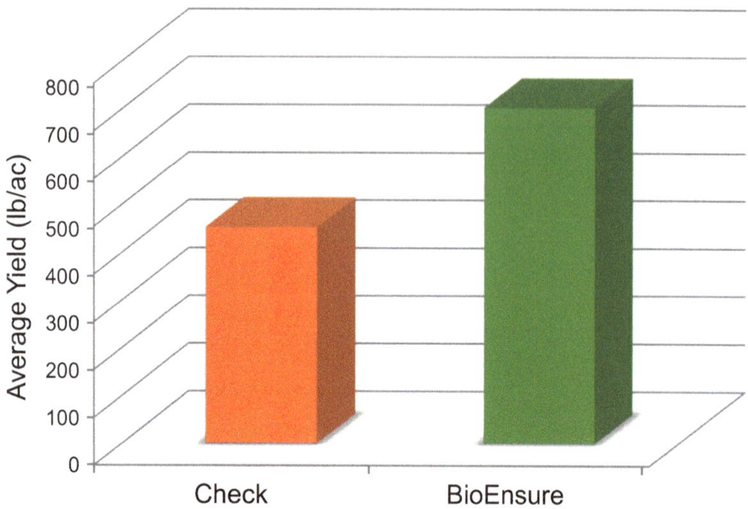

Fig. 5.7 2016 Pearl Millet Field Trial Results with Stress. Studies were conducted in Rajasthan, India, with *BioEnsure*® on mung bean exposed to temperatures up to 45 °C (113 °F) and drought stress. A total of 45 evaluations representing 45 replications, across 45 locations, were conducted by poor rural Indian farmers. The check/control plants are represented in *red* and the *BioEnsure*®-treated plants in *green*. An average yield increase of 53% (250 lb./ac) was observed. Of the 45 evaluations, 96% showed yield increases of ≥1% compared to the check/controls. The seeds were carryover (seed grown by farmer from the previous year) or commercial seeds which were treated with standard PPP, the composition and concentration of which could vary depending upon the seed company and could include the fungicide Thiram, insecticides, polymers, and dyes

5.6 Looking Beyond: A Changing Climate—Multiple Stressors

Facing decreases in agricultural production induced by anthropogenic and climate impacts requires new technologies that can be implemented rapidly to address new and increasing abiotic and biotic stresses. For example, the development of chemical resistance by pests and pathogens, environmental concerns over chemical residues, and changing climate patterns are exacerbating the incidence of disease globally. There is an urgent need to develop microbial products that will provide resistance to both abiotic and biotic stressors. As the threat to global food production increases, there will be elevated levels of political, economic, and social instability followed by massive human migration, much like what is currently happening in the Middle East and Africa. The most expedient approach to these emerging problems is for organizations focusing on different stressors to develop synergistic products that will span the stress continuum. This is an atypical approach to solving problems as organizations tend to work in protective isolation.

In recent years, microbial technologies have also been developed for commercialization by other organizations to address abiotic, biotic and nutritional issues and

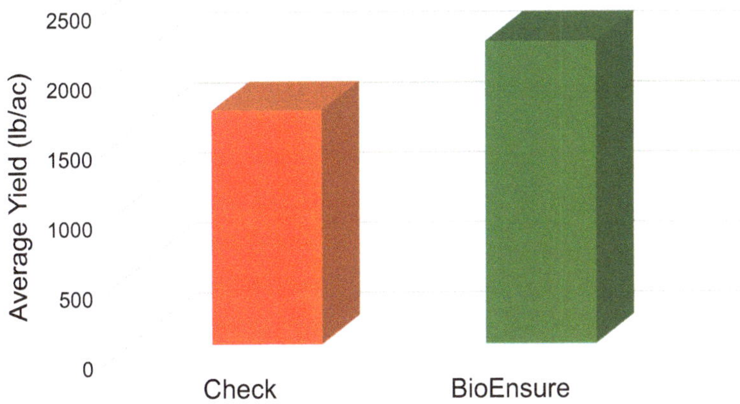

Fig. 5.8 2016 Pearl Millet Field Trial Results with Stress. Studies were conducted in Rajasthan, India, with *BioEnsure*® on pearl millet exposed to elevated high temperatures up to 45 °C (113 °F) and drought stress. A total of 52 evaluations representing 52 replications, across 52 locations, were conducted by poor rural Indian farmers. The check/control plants are represented in *red* and the *BioEnsure*®-treated plants in *green*. An average yield increase of 29% (431 lb./ac) was observed. Of the 52 evaluations, 94% showed yield increases of ≥1% compared to the check/controls. The seeds were carryover (seed grown by farmer from the previous year) or commercial seeds which were treated with standard PPP, the composition and concentration of which could vary depending upon the seed company and could include the fungicide Thiram, insecticides, polymers, and dyes

have been successfully implemented in agriculture. Positive responses both to biotic and abiotic stress, as well as increased yields, have been observed in agricultural plants such as corn, wheat, and rice which are staple food crop plants important for human sustainability and forage for livestock. These results are encouraging as they indicate real-time symbiogenic solutions to big problems.

As the field of symbiosis becomes more embraced by the scientific and agricultural communities, further studies employing combinations of beneficial microbes both symbiotic and rhizospheric along with beneficial insects will provide the complex level of protection required for complex types of stress. To mimic what occurs in native, healthy, stress-tolerant plants, microbial consortia need to be utilized to generate optimal holobiont (host and its microbiota) systems. Such a system would likely offer sufficient genetic potential to address the multitude of challenges that agriculture and society are facing. Further understanding of the biology as well as molecular, physiological, and genetic basis of such complex systems, will aid in the further development of symbiogenic technology for future sustainability.

Acknowledgements We would like to thank our AST team both past and present: Alec Stephen Baird, Alex Marcus Batson, Sang Hyun Cho, Eliza Alston Cliff, Zachery Gray, Evan David Groover, Roman Stentz Harto, Marian Jen-Mei Hsieh, Ryan John Manglona, Malia Marie Mercer, Natalie Marie Nasman, Tatiana Ariane Nicklason, Melissa Nicole Rienstra, and Leesa Lou Wright;

and Mr. Anuj Dayama and Dr. Kalyan Dayama for assistance with organizing field tests in India, interpreting, and logistical support. Funding for field testing in India was supported in part by a Securing Water for Food grant from USAID (AID-OAA-A-15-0007) and a grant from the US-India Science and Technology Endowment Fund (EC-062).

References

1. Bartels D, Sunkar R. Drought and salt tolerance in plants. Crit Rev Plant Sci. 2005;24:23–58.
2. Bray EA, et al. Trends Plant Sci. 1997;2(2):48–54. PubMed PMID: ISI:A1997WG90200004
3. Leone A, Perrotta C, Maresca B. Plant tolerance to heat stress: current strategies and new emergent insight. In: di Toppi LS, Pawlik-Skowronska B, editors. Abiotic stresses in plants. London: Kluwer Academic Pub; 2003. p. 1–22.
4. Redman RS, Sheehan KB, Stout RG, Rodriguez RJ, Henson JM. Thermotolerance conferred to plant host and fungal endophyte during mutualistic symbiosis. Science. 2002;298:1581.
5. Rodriguez RJ, Redman RS, Henson JM. The role of fungal symbioses in the adaptation of plants to high stress environments. Mitig Adapt Strateg Glob Chang. 2004;9:261–72.
6. Smallwood MF, Calvert CM, Bowles DJ. Plant responses to environmental stress. Oxford: BIOS Scientific Publishers Limited; 1999. 224p
7. Wang W, Vincur B, Altman A. Plant responses to drought, salinity and extreme temperatures: towards genetic engineering for stress tolerance. Planta. 2003;218:1–14.
8. Schwaegerle KE. Quantitative genetic analysis of plant growth: biases arising from vegetative propagation. Evol Ecol. 2005;59:1259–67.
9. Yin X, Struik PC, Kropff MJ. Role of crop physiology in predicting gene-to-phenotype relationships. Trends Plant Sci. 2004;9:426–32.
10. Zhou L, Wang JK, Yi Q, Wang YZ, Zhu YG, Zhang ZH. Quantitative trait loci for seedling vigor in rice under field conditions. Field Crops Res. 2007;100:294–301.
11. Devarajan PT, Suryanarayanan TS, Geetha V. Endophytic fungi associated with the tropical seagrass Halophila ovalis (Hydrocharitaceae). Indian J Mar Sci. 2002;31(1):73–4. PubMed PMID: ISI:000179482500012
12. Márquez LM, Redman RS, Rodriguez RJ, Roossinck MJ. A virus in a fungus in a plant – three way symbiosis required for thermal tolerance. Science. 2007;315:513–5.
13. Rodriguez RJ, Henson J, Van Volkenburgh E, Hoy M, Wright L, Beckwith F, et al. Stress tolerance in plants via habitat-adapted symbiosis. ISME J. 2008;2:404–16.
14. Ryan F. Darwin's blind spot: evolution beyond natural selection. Boston: Houghton Mifflin; 2002. 310p
15. Johri BN. Endophytes to the rescue of plants! Curr Sci. 2006;90(10):1315–6. PubMed PMID: ISI:000239381500007
16. Lucero ME, Barrow JR, Osuna P, Reyes I. Plant-fungal interactions in arid and semi-arid ecosystems: large-scale impacts from microscale processes. J Arid Environ. 2006;65(2):276–84. PubMed PMID: ISI:000236645400007
17. Maggio A, Bressan RA, Ruggiero C, Xiong L, Grillo S. Salt tolerance: placing advances in molecular genetics into a physiological and agronomic context. In: di Toppi LS, Pawlik-Skowronska B, editors. Abiotic stresses in plants. London: Kluwer Academic Pub; 2003. p. 53–70.
18. Tuberosa R, Grillo S, Ellis RP. Unravelling the genetic basis of drought tolerance in crops. In: di Toppi LS, Pawlik-Skowronska B, editors. Abiotic stresses in plants. London: Kluwer Academic Pub; 2003. p. 71–122.
19. Bohnert HJ, Nelson DE, Jensen RG. Adaptations to environmental stresses. Plant Cell. 1995;7:1099–111.
20. Jackson AO, Taylor CB. Plant-microbe interactions: life and death at the interface. Plant Cell. 1996;8:1651–68.

21. Macia-Vicente JG, Jansson HB, Talbot NJ, Lopez-Llorca LV. Real-time PCR quantification and live-cell imaging of endophytic colonization of barley (Hordeum vulgare) roots by Fusarium equiseti and Pochonia chlamydosporia. New Phytol. 2009;182(1):213–28. doi:10.1111/j.1469-8137.2008.02743.x. Epub 2009/01/28. PubMed PMID: 19170898
22. Peters S, Dammeyer B, Schulz B. Endophyte-host interactions. I. Plant defense reactions to endophytic and pathogenic fungi. Symbiosis. 1998;25(1–3):193–211. PubMed PMID: ISI:000075581300013
23. Redman RS, Kim YO, Woodward CJ, Greer C, Espino L, Doty SL, et al. Increased fitness of rice plants to abiotic stress via habitat adapted symbiosis: a strategy for mitigating impacts of climate change. PLoS One. 2011;6(7):e14823. doi:10.1371/journal.pone.0014823. Epub 2011/07/14. PubMed PMID: 21750695; PubMed Central PMCID: PMCPMC3130040
24. Stout RG, Al-Niemi TS. Heat-tolerance flowering plants of active geothermal areas in Yellowstone National Park. Ann Bot. 2002;90:259–67.
25. Yuan Z, Druzhinina IS, Labbe J, Redman R, Qin Y, Rodriguez R, et al. Specialized microbiome of a halophyte and its role in helping non-host plants to withstand salinity. Sci Rep. 2016;6:32467. doi:10.1038/srep32467. Epub 2016/08/31. PubMed PMID: 27572178; PubMed Central PMCID: PMCPMC5004162
26. Rodriguez RJ, Redman RS, Henson JM. Symbiotic lifestyle expression by fungal endophytes and the adaptation of plants to stress: unraveling the complexities of intimacy. In: Dighton J, Oudemans P, White J, editors. The fungal community: its organization and role in the ecosystem. Boca Raton: Taylor & Francis/CRC Press; 2005. p. 683–96.
27. Rodriguez RJ, White JFJ, Arnold AE, Redman RS. Fungal endophytes: diversity and functional roles. New Phytol. 2009;182(2):314–30.
28. Schulz B, Boyle C. The endophytic continuum. Mycol Res. 2005;109:661–86. PubMed PMID: ISI:000230687600001
29. Stone JK, Bacon CW, White JF. An overview of endophytic microbes: endophytism defined. In: Bacon CW, White JF, editors. Microbial endophytes. New York: Marcel Dekker, Inc.; 2000. p. 3–30.
30. Bacon CW, Hill NS. Symptomless grass endophytes: products of coevolutionary symbioses and their role in the ecological adaptations of grasses. In: Redkin SC, Carris LM, editors. Endophytic fungi in grasses and woody plants. St. Paul: APS Press; 1996. p. 155–78.
31. Cheplick GP, Clay K. Acquired chemical defenses in grasses – the role of fungal endophytes. Oikos. 1988;52(3):309–18. PubMed PMID: ISI:A1988N893200010
32. Clay K. Fungal endophytes of grasses – a defensive mutualism between plants and fungi. Ecology. 1988;69(1):10–6. PubMed PMID: ISI:A1988L798500003
33. Clay K, Holah J. Fungal endophyte symbiosis and plant diversity in successional fields. Science. 1999;285:1742–5.
34. Schardl C, Leuchtmann A. The Epichloe endophytes of grasses and the symbiotic continuum. In: Dighton J, White JF, Oudemans P, editors. The fungal community: its organization and role in the ecosystem. Boca Raton: Taylor & Francis; 2005. p. 475–503.
35. Schardl SL, Leuchtmann A, Spiering MJ. Symbioses of grasses with seedborne fungal endophytes. Annu Rev Plant Biol. 2004;55:315–40.
36. Maheshwari R. What is an endophytic fungus? Curr Sci. 2006;90(10):1309. PubMed PMID: ISI:000239381500005
37. Murali TS, Suryanarayanan TS, Geeta R. Endophytic Phomopsis species: host range and implications for diversity estimates. Can J Microbiol. 2006;52(7):673–80. PubMed PMID: ISI:000239824600010
38. Redman RS, Dunigan DD, Rodriguez RJ. Fungal symbiosis: from mutualism to parasitism, who controls the outcome, host or invader? New Phytol. 2001;151:705–16.
39. Redman RS, Freeman S, Clifton DR, Morrel J, Brown G, Rodriguez RJ. Biochemical analysis of plant protection afforded by a nonpathogenic endophytic mutant of colletotrichum magna. Plant Physiol. 1999;119:795–804.
40. Redman RS, Rossinck MR, Maher S, Andrews QC, Schneider WL, Rodriguez RJ. Field performance of cucurbit and tomato plants infected with a nonpathogenic mutant of

Colletotrichum magna (teleomorph: *Glomerella magna*; Jenkins and Winstead). Symbiosis. 2002;32:55–70.
41. Arnold AE, Henk DA, Eells RL, Lutzoni F, Vilgalys R. Diversity and phylogenetic affinities of foliar fungal endophytes in loblolly pine inferred by culturing and environmental PCR. Mycologia. 2007;99(2):185–206. m
42. Arnold AE, Lutzoni F. Diversity and host range of foliar fungal endophytes: are tropical leaves biodiversity hotspots? Ecology. 2007;88(3):541–9. PubMed PMID: ISI:000245668500002
43. Arnold AE, Mejia LC, Kyllo D, Rojas EI, Maynard Z, Robbins N, et al. Fungal endophytes limit pathogen damage in a tropical tree. Proc Natl Acad Sci U S A. 2003;100(26):15649–54. PubMed PMID: ISI:000187554600066
44. Druege U, Baltruschat H, Franken P, et al. Sci Hortic. 2007;112(4):422–6. PubMed PMID: CCC:000246430400010
45. Mandyam K, Jumpponen A. Seeking the elusive function of the root-colonising dark septate endophytic fungi. Stud Mycol. 2005;53:173–89. PubMed PMID: ISI:000234257200011
46. Waller F, Achatz B, Baltruschat H, Fodor J, Becker K, Fischer M, et al. The endophytic fungus Piriformospora indica reprograms barley to salt-stress tolerance, disease resistance, and higher yield. Proc Natl Acad Sci U S A. 2005;102:13386–91.
47. Sahay NS, Varma A. *Piriformospora indica*: a new biological hardening tool for micropropagated plants. FEMS Microbiol Lett. 1999;181:297–302.
48. Cline GF, Muth-Spurlock AM, Voelz BE, Lemley CO, Larson JE. Evaluating blood perfusion of the corpus luteum in beef cows during fescue toxicosis. J Anim Sci. 2016;94(1):90–5. doi:10.2527/jas.2015-9554. Epub 2016/01/27. PubMed PMID: 26812315
49. Fayrer-Hosken RA, Hill NS, Heusner GL, Traylor-Wiggins W, Turner K. The effects of ergot alkaloids on the breeding stallion reproductive system. Equine Vet J Suppl. 2013;45:44–7. doi:10.1111/evj.12164. Epub 2013/12/01. PubMed PMID: 24304403
50. Johnson JM, Aiken GE, Phillips TD, Barrett M, Klotz JL, Schrick FN. Steer and pasture responses for a novel endophyte tall fescue developed for the upper transition zone. J Anim Sci. 2012;90(7):2402–9. doi:10.2527/jas.2011-4493. Epub 2012/01/31. PubMed PMID: 22287669
51. Matthews AK, Poore MH, Huntington GB, Green JT. Intake, digestion, and N metabolism in steers fed endophyte-free, ergot alkaloid-producing endophyte-infected, or nonergot alkaloid-producing endophyte-infected fescue hay. J Anim Sci. 2005;83(5):1179–85. Epub 2005/04/14. PubMed PMID: 15827262
52. Schuenemann GM, Edwards JL, Hopkins FM, Rohrbach NR, Adair HS, Scenna FN, et al. Fertility aspects in yearling beef bulls grazing endophyte-infected tall fescue pastures. Reprod Fertil Dev. 2005;17(4):479–86. Epub 2005/05/19. PubMed PMID: 15899160.
53. Shoup LM, Miller LM, Srinivasan M, Ireland FA, Shike DW. Effects of cows grazing toxic endophyte-infected tall fescue or novel endophyte-infected tall fescue in late gestation on cow performance, reproduction, and progeny growth performance and carcass characteristics. J Anim Sci. 2016;94(12):5105–13. doi:10.2527/jas.2016-0819. Epub 2017/01/04. PubMed PMID: 28046145
54. Omacini M, Chaneton EJ, Ghersa CM, Muller CB. Symbiotic fungal endophytes control insect host-parasite interaction webs. Nature. 2001;409(6816):78–81. PubMed PMID: ISI:000166175600042
55. Cheplick GP. Recovery from drought stress in Lolium perenne (Poaceae): are fungal endophytes detrimental? Am J Bot. 2004;91(12):1960–8. doi:10.3732/ajb.91.12.1960. Epub 2004/12/01. PubMed PMID: 21652344
56. Cheplick GP. Persistence of endophytic fungi in cultivars of Lolium perenne grown from seeds stored for 22 years. Am J Bot. 2017;104(4):627–31. doi:10.3732/ajb.1700030. Epub 2017/04/16. PubMed PMID: 28411211
57. Cheplick GP, Harrichandra AP, Liu A. Competitive outcomes depend on host genotype, but not clavicipitaceous fungal endophytes, in Lolium perenne (Poaceae). Am J Bot. 2014;101(12):2068–78. doi:10.3732/ajb.1400264. Epub 2014/12/07. PubMed PMID: 25480704

58. Faeth SH. Are endophytic fungi defensive plant mutualists? (vol 98, pg 25, 2002). Oikos. 2002;99(1):200. PubMed PMID: ISI:000178883100022
59. Saari S, Faeth SH. Hybridization of Neotyphodium endophytes enhances competitive ability of the host grass. New Phytol. 2012;195(1):231–6. doi:10.1111/j.1469-8137.2012.04140.x. Epub 2012/04/12. PubMed PMID: 22489964
60. Saikkonen K, Helander M, Faeth SH, Schulthess F, Wilson D. Endophyte-grass-herbivore interactions: the case of Neotyphodium endophytes in Arizona fescue populations. Oecologia. 1999;121(3):411–20. doi:10.1007/s004420050946. Epub 1999/11/01. PubMed PMID: 28308331
61. Saikkonen K, Phillips TD, Faeth SH, McCulley RL, Saloniemi I, Helander M. Performance of endophyte infected tall fescue in Europe and North America. PLoS One. 2016;11(6):e0157382. doi:10.1371/journal.pone.0157382. Epub 2016/06/11. PubMed PMID: 27284909; PubMed Central PMCID: PMCPMC4902185
62. Rayner MC. Obligate symbiosis in *Calluna vulgaris*. Ann Bot. 1915;29:97–133.
63. Bezerra JD, Nascimento CC, Barbosa Rdo N, da Silva DC, Svedese VM, Silva-Nogueira EB, et al. Endophytic fungi from medicinal plant Bauhinia forficata: diversity and biotechnological potential. Braz J Microbiol. 2015;46(1):49–57. doi:10.1590/S1517-838246120130657. Epub 2015/07/30. PubMed PMID: 26221088; PubMed Central PMCID: PMCPMC4512079
64. Waqas M, Khan AL, Kamran M, Hamayun M, Kang SM, Kim YH, et al. Endophytic fungi produce gibberellins and indoleacetic acid and promotes host-plant growth during stress. Molecules. 2012;17(9):10754–73. doi:10.3390/molecules170910754. Epub 2012/09/11. PubMed PMID: 22960869
65. Weber RW, Stenger E, Meffert A, Hahn M. Brefeldin a production by Phoma medicaginis in dead pre-colonized plant tissue: a strategy for habitat conquest? Mycol Res. 2004;108(Pt 6):662–71. Epub 2004/08/25. PubMed PMID: 15323249
66. Xing YM, Chen J, Cui JL, Chen XM, Guo SX. Antimicrobial activity and biodiversity of endophytic fungi in Dendrobium devonianum and Dendrobium thyrsiflorum from Vietnam. Curr Microbiol. 2011;62(4):1218–24. doi:10.1007/s00284-010-9848-2. Epub 2010/12/25. PubMed PMID: 21181405
67. Petrini O. Taxonomy of endophytic fungi of aerial plant tissues. In: Fokkema NJ, van den Heuvel J, editors. Microbiology of the phyllosphere. Cambridge: Cambridge University Press; 1986. p. 175–87.
68. Pawlowska J, Wilk M, Sliwinska-Wyrzychowska A, Metrak M, Wrzosek M. The diversity of endophytic fungi in the above-ground tissue of two Lycopodium species in Poland. Symbiosis. 2014;63(2):87–97. doi:10.1007/s13199-014-0291-1. Epub 2014/09/30. PubMed PMID: 25264398; PubMed Central PMCID: PMCPMC4174293
69. Redman RS, Litvintseva A, Sheehan KB, Henson JM, Rodriguez R. Fungi from geothermal soils in Yellowstone National Park. Appl Environ Microbiol. 1999;65(12):5193–7. Epub 1999/12/03. PubMed PMID: 10583964; PubMed Central PMCID: PMCPMC91704
70. Redman RS, Sheehan KB, Stout RG, Rodriguez RJ, Henson JM. Thermotolerance generated by plant/fungal symbiosis. Science. 2002;298(5598):1581. doi:10.1126/science.1072191. Epub 2002/11/26. PubMed PMID: 12446900
71. Pirozynski KA, Malloch DW. The origin of land plants a matter of mycotrophism. Biosystems. 1975;6(3):153–64.
72. Remy W, Taylor TN, Hass H, Kerp H. Four hundred-million-year-old vesicular arbuscular mycorrhizae. Proc Natl Acad Sci U S A. 1994;91(25):11841–3.
73. Tu JC. An improved Mathur's medium for growth, sporulation, and germination of spores of *Colletotrichum lindemuthianum*. Microbios. 1985;44(178):87–93.
74. Ricklefs RE. Evolutionary diversification and the origin of the diversity-environment relationship. Ecology. 2006;87:S3–S13.
75. Rodriguez R, Redman R. More than 400 million years of evolution and some plants still can't make it on their own: plant stress tolerance via fungal symbiosis. J Exp Bot. 2008;59(5):1109–14. doi:10.1093/jxb/erm342. Epub 2008/02/13. PubMed PMID: 18267941

Chapter 6
Endophytic Phytohormones and Their Role in Plant Growth Promotion

Shimaila Ali, Trevor C. Charles, and Bernard R. Glick

6.1 Introduction

Microbes can play crucial roles in plants' lives. This relationship varies from beneficial to neutral to pathogenic, to plants [1]. Plant growth-promoting bacteria (PGPB) are those bacteria that provide benefits to the plant and facilitate plant growth in a number of different ways [2]. The mechanisms that bacteria use to promote plant growth can generally be divided into two categories, i.e., direct mechanisms and indirect mechanisms. Direct mechanisms are where bacteria actively and directly either provide the nutrients and other resources that are necessary for plant growth or modulate plant hormonal levels. In indirect mechanisms, bacteria thwart some of the growth inhibition caused by various plant pathogenic agents; the indirect mechanisms include production of lytic enzymes, antibiotics, siderophores, induction of systemic resistance, alteration in ethylene levels, and direct competition with phytopathogens [2].

PGPB that are able to enter and colonize plant tissues without causing them any obvious damage or disease symptoms are termed as bacterial endophytes. Endophytes can typically interact with their hosts more effectively than their plant growth-promoting rhizospheric counterparts [3, 4]. Generally, bacterial endophytes are neither organ nor host specific [5]. A variety of endophytes have been isolated from different tissue types in numerous species of plants hosts [6], and often multiple species of endophytes are found within a single plant [6]. On the other hand, bacteria that are responsible for latent infections and/or colonize senescent plant tissues and produce macroscopic signs of disease are not considered to be

S. Ali (✉)
A&L Biologicals, Agroecological Research Services Centre, London, ON, Canada, N5V 3P5
e-mail: shimaila@alcanada.com; s9rashid@uwaterloo.ca

T.C. Charles • B.R. Glick
Department of Biology, University of Waterloo, Waterloo, ON, Canada, N2L 3G1

S.L. Doty (ed.), *Functional Importance of the Plant Microbiome*,
DOI 10.1007/978-3-319-65897-1_6

endophytes. Endophytic bacteria are also distinguished from bacteria that are present within plant tissues on a transient basis and do not survive there for long periods of time.

It is interesting to note that most laboratory and field PGPB studies primarily have focused on rhizospheric PGPB, while studies of endophytic PGPB have been much more limited. The genetic and biochemical factors that contribute towards successful endophytic colonization and maintenance are not well understood. Nevertheless, there have been a large number of studies on the use of endophytic PGPB as components of various phytoremediation schemes [7]. As a starting point, it should be emphasized that all available evidence indicates that endophytic PGPB employ similar plant growth promotion mechanisms to those used by rhizospheric PGPB [8]. In fact, bacterial endophytes are PGPB that go one step further, i.e., to colonize the inside of the plant tissues where they can serve their host promptly and efficiently compared to those bacteria that dwell exclusively in the plant's rhizosphere. The existing evidence suggests that endophytic PGPB are more effective than similar non-endophytic bacterial strains in promoting plant growth under a wide range of environmental conditions [9, 10].

6.2 Direct Growth Promotion Mechanisms Used by Endophytes

In direct plant growth promotion, bacteria either supply or modulate essential phytohormone levels, including auxin, cytokinin, ethylene, and gibberellin, or provide plants with nutrients such as phosphate, nitrogen, and iron. This chapter focuses on the supply of and modulation of phytohormone levels by endophytes.

6.2.1 Endophyte Hormone Biosynthesis

6.2.1.1 Auxin Biosynthesis

Auxins are plant hormones that play vital roles in almost every step of a plant's daily growth and life [11]. The most common naturally occurring auxin is indole-3-acetic acid (IAA), which is produced by plants and also by bacteria and fungi through at least three different tryptophan-dependent IAA production pathways [12]. The type of pathway a bacterium uses (genetically and/or environmentally dependent) to produce IAA within plants or in their close vicinity can determine the nature of the resulting plant-microbe interactions [13, 14]. It is important to note that not all the IAA-producing bacteria are beneficial to plants. Interestingly, many plant-beneficial bacteria produce IAA via the indole-3-pyruvate (IPyA) pathway, whereas many pathogenic bacteria mainly synthesize IAA via the indole-3-acetamide (IAM) pathway [13].

The areas of activity of IAA mainly include, but are not limited to, cell division and elongation; initiation of root systems, leaves, and flowers; and fruit development and senescence [12, 15]. Numerous PGPB, including both of Gram-positive and Gram-negative bacteria, have been reported to produce IAA [2, 3, 12, 16, 17]. The prominent IAA-producing endophytic bacterial genera include *Pseudomonas, Rhizobium, Azospirillum, Enterobacter, Azotobacter, Klebsiella, Alcaligenes, Pantoea, Acetobacter, Herbaspirillum, Burkholderia, Bacillus, Rhodococcus,* and *Streptomyces* [3, 12, 18, 19]. Generally plants are very sensitive to the amount of IAA present in plant tissue at any particular time. Since plants also produce IAA, in order to regulate plant growth, an IAA-producing PGPB must provide the appropriate amount of IAA (when combined with the amount of the hormone produced by the plant). In fact, phytopathogens are often characterized by their ability to produce IAA at high concentrations [20–23].

Numerous bacterial endophytes have been reported to promote plant growth by their ability to biosynthesize IAA. For example, one study reported IAA production as a common growth promotion trait in bacterial endophytes isolated from apple tree buds, in which 8 of 18 isolates exhibited IAA production of 1.2–2.4 μg/mL [24]. In another study, Vendan et al. [25] investigated the various plant growth-promoting capabilities, including IAA production, of bacterial endophytes from ginseng (*Panax ginseng* C.A. Meyer). A total of 51 bacterial endophytes were isolated from ginseng stem that were clustered in four groups namely *Firmicutes, Actinobacteria*, α-*Proteobacteria*, and γ-*Proteobacteria*, with *Firmicutes* being the most prominent group. Some 18 representatives of all groups were further characterized, and 14 of these 18 endophytic isolates produced significant amounts of IAA when supplemented with tryptophan as a precursor. The highest amount of IAA (13.93 μg/mL) was produced by isolate E-I-4 (*Micrococcus luteus*) and was followed by the isolates E-I-20 (*Lysinibacillus fusiformis*) and E-I-8 (*Bacillus cereus*), which produced 7.23 μg/mL and 4.61 μg/mL, respectively [25, 26]. The population diversity and plant growth promotion effects of IAA produced by endophytic and epiphytic bacteria isolated from soybeans (cultivars Foscarin and Cristalina) have also been investigated [27]. Isolates that presented plant growth promotion capabilities were identified as belonging to genera *Pseudomonas*, *Ralstonia*, *Enterobacter*, *Pantoea*, and *Acinetobacter*. Moreover, endophytic soybean cultivars exhibited more abundant IAA-producing abilities (34%) than epiphytic population (i.e., 21%) [27]. In another study, soybean seed endophytes were isolated from 12 different cultivars of soybeans and identified by amplified ribosomal DNA restriction analysis (ARDRA) grouping and by partial sequencing of their 16S rRNA gene. These endophytes were classified as *Acinetobacter*, *Bacillus*, *Brevibacterium*, *Chryseobacterium*, *Citrobacter*, *Curtobacterium*, *Enterobacter*, *Methylobacterium*, *Microbacterium*, *Micromonospora*, *Pantoea*, *Paenibacillus*, *Pseudomonas*, *Ochrobactrum*, *Streptomyces*, and *Tsukamurella* [28]. They all produced IAA in vitro at significant levels but only one strain (*Enterobacter* sp.) significantly increased the root dry biomass when soybean seeds were pretreated with this strain [28]. Moreover, endophytic bacteria were isolated from crops (berseem clover or canola) in rotation with rice using IAA production as a primary screening trait [29].

This study demonstrated plant growth-promoting features of seven isolates, and rice seedlings inoculated with any one of these isolates exhibited higher shoot biomass, root length, and number of colonizing bacteria than did control plants inoculated with endophytic strains that did not produce IAA [29].

6.2.1.2 Gibberellin Biosynthesis

Gibberellins (GAs) stimulate a number of plant metabolic functions, which are essential for plant growth and development [30] including seed germination, stem elongation, flowering, and fruit formation and senescence [31]. To date, there have been ~136 GAs identified [30–32]. Different GAs are named according to their order of discovery [30]. The full bacterial gibberellin biosynthesis pathway has only recently been described (http://www.nature.com/nchembio/journal/v13/n1/full/nchembio.2232.html) [33]. There is very little known about GA production by bacterial endophytes; only a few studies have described this potential plant growth-promoting trait of bacterial endophytes [34–38]. Two bacterial endophytes, namely *Acetobacter diazotrophicus* and *Herbaspirillum seropedicae*, have been reported to produce gibberellins (GA1 and GA3) and IAA. These bacteria establish an endophytic relationship with graminaceae species where they promote growth and overall yield [35]. In addition, *Azospirillum lipoferum* strain op 33 has been recognized as an endophyte of grasses and a GA producer in vitro [36]. The gibberellins produced by these bacterial endophytes were measured in chemically defined media by capillary gas chromatography-mass spectrometry [35, 36]. In one study [34], spent culture from ten wild-type and mutant strains (including *nod*$^-$ and *fix*$^-$) of *Rhizobium phaseoli* was screened for the presence of GAs and IAA by a high-performance liquid chromatography (HPLC) immunoassay. The presence of GA1, GA4, GA9, and GA20-like molecules was confirmed by combined gas chromatography mass spectrometry; however, the GA20-like molecule was only present in some of the cultures but not in all, and GA9 was detected only in small amounts compared to GA1 and GA4 [34]. The study of *nod*$^-$ and *fix*$^-$ mutants indicated that the production of GAs was independent of genes involved in nodulation and nitrogen fixation in *Rhizobium phaseoli*. More recently, a bacterial endophyte *Sphingomonas* sp. LK11, originally isolated from the leaves of *Tephrosia apollinea* (a legume native to parts of Asia and Africa), has shown the ability, by advanced chromatographic and spectroscopic techniques, to synthesize physiologically active GA4 and inactive GA9 and GA20 in culture media [38]. Tomato plants inoculated with *Sphingomonas* sp. LK11 displayed a significant increase in plant shoot length, chlorophyll contents, and shoot and root dry weights compared to control plants [38].

The other large group of endophytic microorganisms that produces GAs is fungi. A number of fungi have been identified with the capacity to synthesize physiologically active GAs [30, 39–45]. In studies carried out by Hamayun et al. [40, 41], two fungal endophytes, *Aspergillus fumigatus* and *Scolecobasidium tshawytschae*, were isolated from a drought-stressed cultivar (Hwangkeumkong) and a salt-stressed cultivar (Daewonkong) of soybean, respectively. Both strains were identified by morphological characteristics and phylogenetic analysis of 18S ribosomal RNA gene sequences. *Scolecobasidium tshawytschae* produced physiologically active GA1, GA3, GA4, and GA7 and inactive GA15 and GA24, whereas *Aspergillus fumigatus* synthesized active GA3, GA4, and GA7 along with physiologically inactive GA5, GA19, and GA24. Subsequently, rice and soybean plants that were treated with these fungal endophytes show a significant increase in plant length and plant fresh and dry weight compared to plants treated with *Gibberella fujikuroi*, which is also a non-endophytic gibberellin-producing fungus and was used as control for these studies [40, 41]. In addition, the fungal endophyte *Aspergillus fumigatus* sp. LH02 facilitated soybean plant growth under salt stress (70 and 140 mM). The soybean plants pretreated with this fungus exhibited significant increases, compared to the control plants, in shoot length, shoot fresh and dry biomass, leaf area, chlorophyll contents, and photosynthetic rate [42]. It was argued that the treatment of plants with the fungal endophyte, *Aspergillus fumigatus* sp. LH02, increased the plant's levels of proline, salicylic acid (SA), and jasmonic acid (JA) and lowered the abscisic acid (ABA) concentration compared to control (uninoculated plants) [42]. Moreover, the treated plants had higher levels of isoflavones, which was independent of the level of salt stress, notwithstanding the fact that isoflavones are considered to be a factor in helping soybean plants to cope with salt stress [42]. Similarly, a group of 11 fungal endophytes was isolated from the sand-dune plant *Elymus mollis* and subsequently screened for the production of GAs and their plant growth-promoting capacities on Waito-c rice (GA-deficient rice) and *Atriplex gemelinii* (saltbush). Altogether, 7 of 11 fungal endophytes promoted the growth of both plants, and one isolate, EM-7-1, showed significantly higher plant growth compared to control plants [39]. Screening of the culture filtrate of isolate EM-7-1 revealed the presence of GA1, GA3, GA4, and GA7 as well as physiologically inactive GA5, GA9, GA20, and GA24. Subsequently, this isolate was identified as *Gliomastix murorum* [39]. A similar study was reported by Khan et al. [43] who isolated and characterized two fungal endophytes from the bark of *Moringa peregrine* (a tree indigenous to the Horn of Africa), *Aspergillus caespitosus* LK12 and *Phoma* sp. LK13, and showed that these fungi could produce a variety of GAs in culture filtrate. Both fungal strains promoted the growth of rice that lacks in gibberellin biosynthesis [43]. Furthermore, two GA- and IAA-producing fungal endophytes *Phoma glomerat*a LWL2 and *Penicillium* sp. LWL3 have been shown to provide protection to cucumber plants under salinity and drought stresses [45]. The treated plants exhibited significantly higher contents of a range of nutrients than the untreated control plants [45]. On the other hand, under salinity stress, treated cucumber plants upregulated SA levels, altered JA levels, and downregulated ABA levels and glutathione, catalase, peroxidase, and polyphenol oxidase activities. This change in the metabolome of the treated cucumber plants compared to uninoculated control plants is ascribed to the endophytic fungus, which apparently ameliorated the detrimental effects of the stress [45]. The effects of GA-producing strains of *Penicillium* sp. on plants under salt stress were evaluated by Leitão and Enguita [44], who described the role of the fungus in overcoming the stress incurred by salinity [44].

6.2.1.3 Cytokinin Biosynthesis

Cytokinins are a group of hormones that promote cell division in plant roots, shoots, and growing buds. These hormones have been found in all complex plants as well as mosses, fungi, and bacteria. There are about 200 different natural and synthetic cytokinins known to botanists today. Most cytokinins are produced in the meristem of the roots and transported to the other parts of the plant through the xylem (vascular system) [46]. Cytokinins have been reported to be present in the culture filtrate of a number of bacteria including *Azotobacter* sp., *Rhizobium* sp., *Pantoea agglomerans*, *Rhodospirillum rubrum*, *Pseudomonas fluorescens*, *Bacillus subtilis*, and *Paenibacillus polymyxa* [2]. Although these bacteria have been documented as endophytes of different plants, little evidence has been found to definitively link bacterial cytokinin production with plant growth promotion. One study described the isolation, identification, and characterization of bacterial endophytes that produce cytokinin-like molecules [47]. In this study, three bacterial endophytes were isolated from Sambung Nyawa (*Gynura procumbens* (Lour.) Merr.) and identified as *Pseudomonas resinovorans*, *Paenibacillus polymyxa*, and *Acinetobacter calcoaceticus*. The ethyl acetate extract of bacterial culture media was used to inoculate cucumber cotyledons in a greening bioassay. The assay indicated positive results only for the strains *Pseudomonas resinovorans* and *Paenibacillus polymyxa* with the suggestion that these bacterial endophytes might be used as plant growth-promoting agents for Sambung Nyawa [47]. In a similar study, crude bacterial suspensions of 115 bacterial isolates from 72 different plant species were screened using the cucumber cotyledon greening bioassay to investigate if these endophytes could produce cytokinins, but the study found that none yielded better results than the control [48]. Recently, workers engineered a strain of *Sinorhizobium meliloti* to overproduce cytokinin by expressing an *Agrobacterium ipt* gene under the control of the *E. coli trp* promoter [49]. Following a period of severe drought stress, alfalfa plants inoculated with the engineered *S. meliloti* strain were significantly larger than plants inoculated with the parental strain. This experiment indicates, despite the fact cytokinin-producing plant growth-promoting bacteria appear to be relatively uncommon, that rhizobial strains synthesizing higher than normal levels of cytokinin may improve plant tolerance to severe drought stress.

6.2.2 Endophytic ACC Deaminase Production

The microbial enzyme 1-aminocyclopropane 1-carboxylate (ACC) deaminase (E.C. 4.1.99.4) has been studied by a large number of researchers including, but not limited to [3, 13, 50–63]; and [64]. These researchers have discussed the enzymology, biochemistry, mechanism, regulation, and potential use of this dynamic enzyme in sustainable agriculture. Concisely, ACC deaminase is a multimeric enzyme that requires pyridoxal 5′-phosphate as an essential cofactor for enzymatic activity [50, 65] and cleaves ACC to α-ketobutyrate and ammonia, where ACC is the immediate precursor of the phytohormone ethylene. In this way, ACC deaminase lowers the levels of deleterious ethylene in higher plants [3, 59, 66–70]. Ethylene, like other phytohormones, is crucial for plant growth, development, and stress signaling [3, 71–74]. Plant growth-promoting endophytes expressing the enzyme ACC deaminase have been shown to help protect plants from a number of different biotic and abiotic stresses and also promoted the growth of plants in the absence of stressful conditions. For example, a group of 25 endophytes, originally isolated from tomato plants, that contain ACC deaminase all demonstrated the ability to significantly promote canola seedling growth compared to uninoculated canola seedlings [3]. Different studies have described the potential of ACC deaminase containing endophytes in promoting plant growth in tomato [75–77], rice [78], and ginger [79]. Recently [80], the potential of the bacterial endophyte *Burkholderia phytofirmans* PsJN was evaluated in a field study of switchgrass in two different soil sites over 2 years. The inoculated switchgrass displayed enhanced biomass production, increased root growth, tillering, and greater early-season plant growth vigor than the untreated control plants. Moreover, the plants grown on a low-fertility soil site performed better with the bacterial endophyte treatment. These researchers suggested that the mechanism of this plant growth promotion, especially at the poor soil site, might include the possession of ACC deaminase by this endophyte and the interaction of auxins and ethylene in response to the action of ACC deaminase [80]. In addition to this work, it has been previously shown that a mutant of *Burkholderia phytofirmans* PsJN that lacks the ability to produce ACC deaminase (i.e. $acdS^{-}$) could not promote canola seedling growth in a growth pouch root elongation assay [70]. Complementing the mutant with exogenous DNA carrying the ACC deaminase gene from the wild-type strain restored the ability of the mutant to promote canola root elongation, therefore proving the importance of this gene for the observed growth promotion [70]. More recently, it was found that *B. phytofirmans* PsJN specifically promoted the growth of a certain genotype of switchgrass, which led to the isolation and characterization of another bacterial endophyte, *Pantoea agglomerans* strain PaKM, from the surface of sterilized seeds of switchgrass [81]. Strain PaKM was able to promote the growth of at least eight varieties of switchgrass in in vitro conditions; subsequently two of these varieties were screened more extensively in the greenhouse and in the field environment, and a significant difference in the biomass of the endophytic treated plants was observed [81]. *P. agglomerans* strain PaKM was also able to protect switchgrass under salt and drought stress in in vitro conditions; however, this endophyte does not contain ACC deaminase [81].

When the bacterial endophyte *Pseudomonas migulae* 8R6, an ACC deaminase-containing bacterium, was utilized as a biocontrol agent against yellow disease of grapevines caused by phytoplasma, it significantly protected periwinkle, a model plant hosting phytoplasma. The results have shown that the density of the phytoplasma inside the leaf tissue was unaffected by this bacterial endophyte; however, the symptoms of the disease were significantly reduced in the plants treated with the wild-type bacterium compared with the plants either untreated or treated with an ACC deaminase minus mutant ($acdS^{-}$) of strain 8R6 [82]. These experiments suggest that the ACC deaminase played a key role in protecting plant from the biotic stress of phytoplasma infection. Moreover, ACC deaminase-containing endophytes have also been found to protect plants from salinity and other abiotic stress. The rice endophyte *Pseudomonas stutzeri* A1501 has demonstrated rice seedling growth promotion in moderate (0.12 M) and high (2 M) salt (i.e., NaCl) and in the presence of 0.3 mM heavy metals (i.e., Cu, Co, Ni, and Zn) [83]. In the above-mentioned experiments, the bacterial treatment was given to surface-sterilized rice seeds and the plant biometrics were collected after 7 days. In order to validate that the ACC deaminase activity is the main driving force in protecting and facilitating rice seedling growth in the presence of these abiotic stresses, a mutant of the *acdS* gene was constructed. It was observed that the seeds treated with wild-type *P. stutzeri* A1501 displayed significantly longer roots and higher fresh and dry weights compared to the plants either untreated or treated with the mutant [83]. In a similar study, endophytic bacteria isolated from date palm were assayed for growth promotion of canola roots in the presence and absence of 100 mM salt. The majority of the endophytes tested in this study exhibited canola root elongation under salt stress compared to uninoculated control plants in gnotobiotic conditions; however, the researchers pointed out that these endophytes, in addition to ACC deaminase, could also produce IAA and increase the uptake of nutrients that enable them to benefit the host under stress [84]. *Brachybacterium paraconglomeratum* is an ACC deaminase-producing salt-tolerant bacterial endophyte, which was isolated from the surface-sterilized roots of a medicinal plant, *Chlorophytum borivilianum* [85]. This bacterium promoted host plant growth by reducing oxidative and osmotic damages caused by salinity (150 mM). Moreover, biochemical analysis of bacterially treated and untreated plants, both grown in the presence of salt, revealed that there were high amounts of ACC, proline, malondialdehyde (MDA), and abscisic acid (ABA) found in the untreated control plants. Increased levels of proline and MDA indicate osmotic and oxidative stress, respectively, whereas increased ABA and ACC levels are thought to be the consequence of osmo-oxidative damage [85]. However, plants treated with the wild-type endophyte (containing ACC deaminase) show reduced levels of proline, MDA, ABA, and ACC, and increased total chlorophyll contents, IAA levels, and plant biomass compared to untreated control plants [85].

Additionally, the expression of five genes involved in the stress response (i.e., *CaACCO*, *CaLTPI*, *CaSAR82A*, and putative *P5CR* and *P5CS*) in pepper plants (*Capsicum annuum* L.) was investigated. These plants were given a mild osmotic stress in the presence and absence of two plant growth-promoting bacterial endophytes, *Arthrobacter* spp. EZB4 and *Bacillus* spp. EZB8 [86]. The pepper plant gene, *CaACCO*, which encodes the enzyme ACC oxidase that catalyzes the final step in the biosynthesis of ethylene, was strongly upregulated in noninoculated stressed plant root and leaf tissue. This gene was significantly less upregulated in leaf tissue, unaffected in plant roots treated with strain EZB4, and unaffected in the leaf and root tissue of plants treated with strain EZB8. The pepper plant gene *CaLTPI* encodes a lipid transfer protein that may be induced by ethylene in addition to a number of other stress factors

[87]. The gene expression of *CaLTPI* was significantly upregulated under osmotic stress in noninoculated plants, unaltered in plants (leaves and roots) treated with strain EZB4 and the leaf tissue of plants treated with strain EZB8, and significantly downregulated in the roots treated with strain EZB8. The putative *P5CR* and *P5CS* are involved in proline biosynthesis and were significantly downregulated in the leaves of plants treated with both endophytic strains under stress conditions and remained unaffected in the roots of stressed plants. The expression of the pepper plant gene *CaSAR82A*, which is also a stress-inducible gene, was not consistent under stress conditions in either leaf or root tissues with bacterial treatments, but was significantly upregulated in noninoculated stressed plants [86]. Altogether, it was speculated that because ethylene can act as signaling molecule and subsequently could regulate the gene expression under stress conditions, the addition of endophytes that can lower stress ethylene levels by the functioning of ACC deaminase could ameliorate the damage caused by such stress. Nonetheless, independent of the levels of altered gene expression of the above-mentioned genes, all of the pepper plants treated with the bacterial endophytes *Arthrobacter* spp. EZB4 and *Bacillus* spp. EZB8 showed significant increases in biomass compared to the noninoculated control plants, under mild osmotic stress [86].

In addition to plant growth facilitation under stressful environmental conditions, ACC deaminase has also been documented to help endophytic colonization within plants [13]. Ethylene levels in plant tissues modulate plant colonization by endophytes [13, 88]. The bacterial endophyte *Klebsiella pneumoniae* strain 342 can establish endophytic relationships with *Medicago truncatula*; however, the colonization of plants by this endophyte was found to be under the control of ethylene. In an ethylene-insensitive mutant of *Medicago truncatula*, this endophyte hypercolonizes the plant compared to the wild-type [88] *Medicago truncatula*, which displayed a low level of endophytic colonization in the presence of ACC, and an increase in endophytic colonization was observed when the ethylene inhibitor, 1-methylcyclopropene, was introduced to the plant [88]. Since ACC deaminase is able to lower the levels of ethylene by cleaving ACC, it may help some endophytes to efficiently colonize plant tissues, thereby giving those endophytes an additional advantage in their interaction with plants.

6.3 A Model of Endophytic Plant Growth Promotion

A detailed model of some of the key aspects of plant growth promotion by PGPB has been presented [50, 51, 67]. That model (Fig. 6.1), which is essentially the same for both endophytes and rhizosphere bacteria, presents the mechanism that an ACC deaminase and IAA-producing bacterium uses in order to facilitate plant growth, especially under abiotic or biotic stress conditions. These stresses include salinity, drought, flooding, soil toxicity with heavy metals and organics, and various bacterial and fungal pathogens. Endophytes are better adapted within

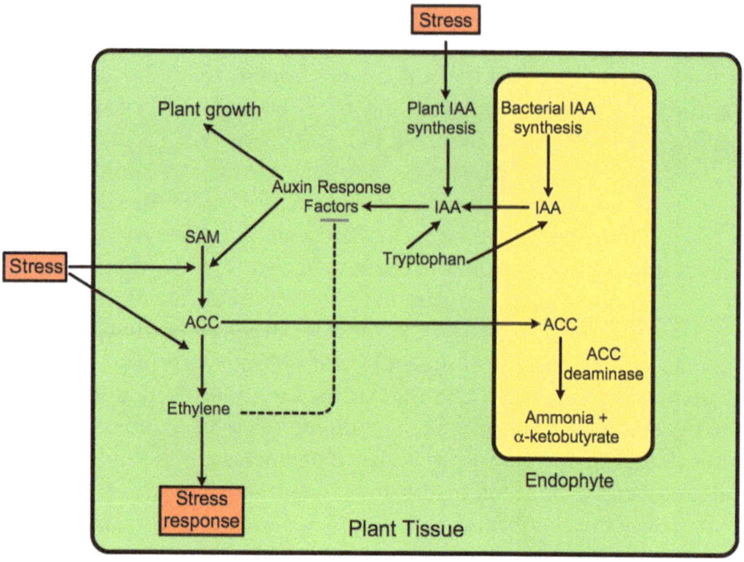

Fig. 6.1 A proposed model of plant growth promotion by endophytes that possess ACC deaminase and are able to produce IAA within a host plant. This figure shows a plant tissue harboring an endophyte and the interactions between them

the plant environment and interact with plants quicker and more promptly over their rhizospheric counterparts [3, 13]. As discussed earlier, ACC deaminase is an inducible enzyme and is expressed only at a low level in the absence of its substrate and inducer, ACC. Being inside of plant tissue, an ACC deaminase-expressing endophyte will have the following: it can (1) take up and utilize small nutrient molecules that are synthesized by the plant, (2) be easily affected by the stress a plant may face, and (3) help to protect its host plant immediately following a stress and even before the onset of damaging symptoms that occur as a consequence of such a stress. The increased levels of IAA, which can be produced and transported to plant tissues by residential endophytes, in addition to promoting plant growth, also trigger the synthesis of ACC. This occurs because IAA induces the transcription of the mRNA encoding the plant enzyme ACC synthase that converts *S*-adenosyl methionine (SAM) to ACC [50]. The increased levels of ACC induce the synthesis of the enzyme ACC deaminase in the endophyte, and in response, a large portion of the ACC (but not all of it) is broken down to ammonia and α-ketobutyrate. In this way, the ACC levels (either the plant endogenous ACC or the IAA-induced ACC, or a combination of both) within the plant tissue are decreased. Thus, endophytes that contain ACC deaminase rescue plants from producing deleterious ethylene levels [89]. Since ACC oxidase, which is responsible for the biosynthesis of ethylene in the plant, has a lower K_m value for ACC than ACC deaminase [51, 67], it is imperative that ACC deaminase function before any significant induction of plant ACC oxidase occurs;

otherwise plant inhibitory levels of ethylene will result [50]. The interaction between ACC deaminase, IAA production, and stress response factors has been described in a microarray study of the ACC deaminase-producing rhizospheric bacterium *P. putida* UW4 inoculated onto canola seeds [63]. In this study it was shown that wild-type *P. putida* UW4 upregulated plant genes involved in auxin signal transduction and downregulated plant genes that encode proteins that protect the plant from a variety of environmental stresses, since these latter proteins are no longer needed under lowered ethylene concentrations [50]. In summary, this model predicts that a bacterium that produces both ACC deaminase and IAA promotes plant growth by dint of lowering ethylene levels in the plant and thereby allowing a greater flux of IAA signal transduction; therefore growth promotion per se is a consequence of IAA functioning.

6.4 Hormonal Based Endophytic Environmental Stress Control

Since the increased levels of ethylene could lead plants to enter stress, the endophytes that express ACC deaminase have been known to help host plants to overcome environmental stresses that they face over the span of their life. Below are some examples of such interactions.

6.4.1 Flower Wilting

Flowers undergo senescence events as part of their natural growth cycle. Bacterial endophytes have been shown to be able to prolong flowers' shelf life [10]. In this study, two bacterial endophytes belonging to the genus *Pseudomonas* were used to evaluate their effects on mini carnation flower wilting. Both of the endophytes express high levels of ACC deaminase, produce IAA and siderophores, and solubilize inorganic phosphate in vitro [3]. Carnation flowers undergo rapid senescence after the production of high levels of ethylene and hence are considered as ethylene-sensitive plants [90]. ACC deaminase-expressing bacterial endophytes can provide protection to carnation cut flowers preventing the production of stress ethylene by lowering ACC concentrations in the tissues and thereby delaying flower senescence for 2–3 additional days over untreated control flowers or flowers treated with *acdS*-deficient bacterial mutants [10]. Commercially, many cut flowers, including carnations, are treated with a chemical inhibitor of ethylene (i.e., silver thiosulfate) to extend the flowers' shelf life [91]; therefore, the environmentally friendly use of nonhazardous bacterial endophytes could be an environmentally attractive alternative to the use of chemicals in the multibillion dollar cut flower industry [10, 50].

6.4.2 Salinity Stress

Salinity is one of the major growing problems that the agriculture sector is facing. Bacterial endophytes have been successfully used to protect a variety of plants under salt stress [9, 83, 85]. Plants under salinity stress typically produce higher levels of ethylene, which can be prevented by the action of the enzyme ACC deaminase [50]. Two bacterial endophytes *Pseudomonas fluorescens* YsS6 and *Pseudomonas migulae* 8R6, isolated from tomato plants, can provide protection and growth promotion to tomatoes under a very high level (185 mM) of salt stress [9]. These bacterial endophytes produce ACC deaminase, IAA, siderophores, and solubilize phosphate in vitro [3]. In order to eliminate the growth-promoting contributions from other traits, ACC deaminase-deficient mutants were constructed for both endophytes. Surface-sterilized tomato seeds were incubated with wild-type or mutant endophytes, or no added bacteria (as a control). Under salinity stress, plants treated with wild-type bacterial endophytes illustrated significantly higher chlorophyll contents, higher biomass, more fruits, and lower sodium contents compared to plants treated with ACC deaminase minus mutants or no bacteria [9]. This work, along with the other studies presented here, clearly demonstrates the ability of bacterial endophytes containing ACC deaminase to facilitate plant growth under ethylene-induced stress conditions.

6.5 Summary and Conclusions

Plants harbor and interact with a variety of microbial populations at various stages of their lifetime; these dynamic interactions may be harmful, benign, or beneficial to plants. Endophytism is a mutualistic plant-microbe interaction where plants provide a safe home and secure supply of food to microbes, and microbes, in return, benefit the plants enormously. A number of researchers have studied these beneficial interactions where endophytes have been shown to provide nutrients and growth regulators, fixed nitrogen, antibiotics, and other secondary metabolites, and overall the ability to protect plants from both abiotic and biotic stresses. Moreover, endophytes are thought to be superior to their rhizospheric counterparts in facilitating plant growth as they have the ability to colonize the interior of plants where they can rapidly sense any changing environment and quickly respond to their own and the plant's needs. Since endophytes are generally not host specific, endophytes with a desired set of plant growth-promoting activities can be readily introduced into plants other than their natural host. In addition, a number of attempts have been made to genetically engineer bacterial endophytes in an effort to make them more efficient plant colonizers and better plant growth promoters [7]. In light of these considerations, the use of endophytes in agriculture can offer an economic means of achieving high crop productivity and hence sustainable agriculture.

References

1. Lynch JM, editor. The rhizoshere. Chichester, UK: Wiley-Interscience; 1990.
2. Glick BR. Plant growth-promoting bacteria: mechanisms and applications. Scientifica. 2012;2012:963401.
3. Rashid S, Charles TC, Glick BR. Isolation and characterization of new plant growth-promoting bacterial endophytes. Appl Soil Ecol. 2012;61:217–24.
4. Reiter B, Sessitsch A. Bacterial endophytes of the wildflower *Crocus albiflorus* analyzed by characterization of isolates and by a cultivation-independent approach. Can J Microbiol. 2006;52(2):140–9.
5. Rosenblueth M, Martínez-Romero E. Bacterial endophytes and their interactions with hosts. Mol Plant-Microbe Interact. 2006;19(8):827–37.
6. Kobayashi DY, Palumbo JD. Bacterial endophytes and their effects on plants and uses in agriculture. In: Bacon CW, White JF, editors. Microbial endophytes. New York: Marcel Dekker, Inc.; 2000. p. 199–233.
7. Doty SL. Enhancing phytoremediation through the use of transgenics and endophytes. New Phytol. 2008;179(2):318–33.
8. Santoyo G, Moreno-Hagelsieb G, del Carmen Orozco-Mosqueda M, Glick BR. Plant growth-promoting bacterial endophytes. Microbiol Res. 2016;183:92–9.
9. Ali S, Charles TC, Glick BR. Amelioration of high salinity stress damage by plant growth-promoting bacterial endophytes that contain ACC deaminase. Plant Physiol Biochem. 2014a;80:160–7.
10. Ali S, Charles TC, Glick BR. Delay of flower senescence by bacterial endophytes expressing 1-aminocyclopropane-1-carboxylate deaminase. J Appl Microbiol. 2012;113(5):1139–44.
11. Grossmann K. Auxin herbicides: current status of mechanism and mode of action. Pest Manag Sci. 2010;66(2):113–20.
12. Duca D, Lorv J, Patten CL, Rose D, Glick BR. Indole-3-acetic acid in plant-microbe interactions. Antonie Van Leeuwenhoek. 2014;106(1):85–125.
13. Hardoim PR, van Overbeek LS, Elsas JDv. Properties of bacterial endophytes and their proposed role in plant growth. Trends Microbiol. 2008;16(10):463–71.
14. Spaepen S, Vanderleyden J, Remans R. Indole-3-acetic acid in microbial and microorganism-plant signaling. FEMS Microbiol Rev. 2007;31(4):425–48.
15. Phillips KA, Skirpan AL, Liu X, Christensen A, Slewinski TL, Hudson C, Barazesh S, Cohen JD, Malcomber S, McSteen P. Vanishing tassel2 encodes a grass-specific tryptophan aminotransferase required for vegetative and reproductive development in maize. Plant Cell. 2011;23(2):550–66.
16. Glick BR. The enhancement of plant growth by free-living bacteria. Can J Microbiol. 1995;41(2):109–17.
17. Patten CL, Glick BR. Regulation of indoleacetic acid production in *Pseudomonas putida* GR12-2 by tryptophan and the stationary-phase sigma factor RpoS. Can J Microbiol. 2002;48(7):635–42.
18. Apine OA, Jadhav JP. Optimization of medium for indole-3-acetic acid production using *Pantoea agglomerans* strain PVM. J Appl Microbiol. 2011;110(5):1235–44.
19. Costacurta A, Vanderleyden J. Synthesis of phytohormones by plant-associated bacteria. Crit Rev Microbiol. 1995;21(1):1–18.
20. Davies PJ. Plant hormones: physiology, biochemistry, and molecular biology. Boston: Kluwer Academic; 1995.
21. Kunkel BN, Chen Z. Virulence strategies of plant pathogenic bacteria. In: Dworkin M, editor. The prokaryotes. A handbook on biology of bacteria, ecophysiology and biochemistry. New York: Springer; 2006. p. 421–40.
22. Rezzonico E, Flury N, Meins F Jr, Beffa R. Transcriptional down-regulation by abscisic acid of pathogenesis-related β-1,3-glucanase genes in tobacco cell cultures. Plant Physiol. 1998;117(2):585–92.

23. Shinshi H, Mohnen D, Meins F Jr. Regulation of a plant pathogenesis-related enzyme: inhibition of chitinase and chitinase mRNA accumulation in cultured tobacco tissues by auxin and cytokinin. Proc Natl Acad Sci U S A. 1987;84(1):89–93.
24. Miliūtė I, Buzaitė O. IAA production and other plant growth promoting traits of endophytic bacteria from apple tree. Biologija. 2011;57(2):98–102.
25. Vendan RT, Yu YJ, Lee SH, Rhee YH. Diversity of endophytic bacteria in ginseng and their potential for plant growth promotion. J Microbiol. 2010;48(5):559–65.
26. Lacuna PT, Azevedo JL, editors. Endophytic bacteria: a biotechnological potential in agrobiology system. Berlin: Springer; 2013.
27. Kuklinsky-Sobral J, Araújo WL, Mendes R, Geraldi IO, Pizzirani-Kleiner AA, Azevedo JL. Isolation and characterization of soybean-associated bacteria and their potential for plant growth promotion. Environ Microbiol. 2004;6(12):1244–51.
28. Assumpção LDC, Lacava PT, Dias AC, de Azevedo JL, Menten JOM. Diversity and biotechnological potential of endophytic bacterial community of soybean seeds. Pesqui Agropecu Bras. 2009;44(5):503–10.
29. Etesami H, Alikhani HA, Hosseini HM. Indole-3-acetic acid (IAA) production trait, a useful screening to select endophytic and rhizosphere competent bacteria for rice growth promoting agents. MethodsX. 2015;2:72–8.
30. Khan AL, Hussain J, Al-Harrasi A, Al-Rawahi A, Lee I. Endophytic fungi: resource for gibberellins and crop abiotic stress resistance. Crit Rev Biotechnol. 2015;35(1):62–74.
31. Bömke C, Tudzynski B. Diversity, regulation, and evolution of the gibberellin biosynthetic pathway in fungi compared to plants and bacteria. Phytochemistry. 2009;70(15–16):1876–93.
32. Hedden P, Phillips AL, Rojas MC. Gibberellin biosynthesis in plants and fungi: a case of convergent evolution? J Plant Growth Regul. 2002;20:319–31.
33. Nett RS, Montanares M, Marcassa A, Lu X, Nagel R, Charles TC, Hedden P, Rojas MC, Peters RJ. Elucidation of gibberellin biosynthesis in bacteria reveals convergent evolution. Nat Chem Biol. 2017;13(1):69–74.
34. Atzorn R, Crozier A, Wheeler CT, Sandberg G. Production of gibberellins and indole-3-acetic acid by *Rhizobium phaseoli* in relation to nodulation of *Phaseolus vulgaris* roots. Planta. 1988;175(4):532–8.
35. Bastián F, Cohen A, Piccoli P, Luna V, Baraldi R, Bottini R. Production of indole-3-acetic acid and gibberellins A1 and A3 by *Acetobacter diazotrophicus* and *Herbaspirillum seropedicae* in chemically-defined culture media. Plant Growth Regul. 1998;24(1):7–11.
36. Bottini R, Fulchieri M, Pearce D, Pharis RP. Identification of gibberellins A1, A3, and Iso-A3 in cultures of *Azospirillum lipoferum*. Plant Physiol. 1989;90(1):45–7.
37. Crozier A, Arruda P, Jasmin J, Monteiro AM, Sandberg G. Analysis of Indole-3-acetic acid and related indoles in culture medium from *Azospirillum lipoferum* and *Azospirillum brasilense*. Appl Environ Microbiol. 1988;54:2833–7.
38. Khan AL, Waqas M, Kang S, Al-Harrasi A, Hussain J, Al-Rawahi A, Al-Khiziri S, Ullah I, Ali L, Jung H, Lee I. Bacterial endophyte Sphingomonas sp. LK11 produces gibberellins and IAA and promotes tomato plant growth. J Microbiol. 2014b;52(8):689–95.
39. Afzal Khan S, Hamayun M, Kim H, Yoon H, Lee I, Kim J. Gibberellin production and plant growth promotion by a newly isolated strain of *Gliomastix murorum*. World J Microbiol Biotechnol. 2009;25(5):829–33.
40. Hamayun M, Khan SA, Khan MA, Khan AL, Kang S, Kim S, Joo G, Lee I. Gibberellin production by pure cultures of a new strain of *Aspergillus fumigatus*. World J Microbiol Biotechnol. 2009a;25(10):1785–92.
41. Hamayun M, Khan SA, Kim H, Chaudhary MF, Hwang Y, Shin D, Kim I, Lee B, Lee I. Gibberellin production and plant growth enhancement by newly isolated strain of *Scolecobasidium tshawytschae*. J Microbiol Biotechnol. 2009b;19(6):560–5.
42. Khan AL, Hamayun M, Kim Y, Kang S, Lee J, Lee I. Gibberellins producing endophytic *Aspergillus fumigatus* sp. LH02 influenced endogenous phytohormonal levels, isoflavonoids production and plant growth in salinity stress. Process Biochem. 2011;46(2):440–7.

43. Khan AL, Waqas M, Hussain J, Al-Harrasi A, Al-Rawahi A, Al-Hosni K, Kim M, Adnan M, Lee I. Endophytes *Aspergillus caespitosus* LK12 and *Phoma* sp. LK13 of *Moringa peregrina* produce gibberellins and improve rice plant growth. J Plant Interact. 2014a;9(1):731–7.
44. Leitão AL, Enguita FJ. Gibberellins in *Penicillium* strains: challenges for endophyte-plant host interactions under salinity stress. Microbiol Res. 2016;183:8–18.
45. Waqas M, Khan AL, Kamran M, Hamayun M, Kang S, Kim Y, Lee I. Endophytic fungi produce gibberellins and indoleacetic acid and promotes host-plant growth during stress. Molecules. 2012;17(9):10754–73.
46. Greene EM. Cytokinin production by microorganisms. Bot Rev. 1980;46(1):25–74.
47. Bhore SJ, Nithaya R, Loh CY. Screening of endophytic bacteria isolated from leaves of Sambung Nyawa [*Gynura procumbens* (Lour.) Merr.] for cytokinin-like compounds. Bioinformation. 2010;5(5):191–7.
48. Bhore SJ, Sathisha G. Screening of endophytic colonizing bacteria for cytokinin-like compounds: crude cell-free broth of endophytic colonizing bacteria is unsuitable in cucumber cotyledon bioassay. World J Agric Sci. 2010;6(4):345–52.
49. Xu J, Li X, Luo L. Effects of engineered *Sinorhizobium meliloti* on cytokinin synthesis and tolerance of alfalfa to extreme drought stress. Appl Environ Microbiol. 2012;78(22):8056–61.
50. Glick BR. Bacteria with ACC deaminase can promote plant growth and help to feed the world. Microbiol Res. 2014;169(1):30–9.
51. Glick BR, Todorovic B, Czarny J, Cheng Z, Duan J, McConkey B. Promotion of plant growth by bacterial ACC deaminase. Crit Rev Plant Sci. 2007;26(5–6):227–42.
52. Honma M. Chemically reactive sulfhydryl groups of 1-aminocyclopropane-1-carboxylate deaminase. Agric Biol Chem. 1985;49(3):567–71.
53. Honma M. Enzymatic determination of 1-aminocyclopropane-1-carboxylic acid. Agric Biol Chem. 1983;47(3):617–8.
54. Honma M, Smmomura T. Metabolism of 1-aminocyclopropane-1-carboxylic acid. Agric Biol Chem. 1978;42(10):1825–31.
55. Honma M, Kawai J, Yamada M. Identification of the reactive sulfhydryl group of 1-aminocyc lopropane-1-carboxylate deaminase. Biosci Biotechnol Biochem. 1993a;57(12):2090–3.
56. Honma M, Kirihata M, Uchimura Y, Ichimoto I. Enzymatic deamination of (±)-2-methyl- and (s)-2,2-dimethyl-1-aminocyclopropane-1-carboxylic acid. Biosci Biotechnol Biochem. 1993b;57(4):659–61.
57. Hontzeas N, Saleh SS, Glick BR. Changes in gene expression in canola roots induced by ACC-deaminase- containing plant-growth-promoting bacteria. Mol Plant-Microbe Interact. 2004a;17(8):865–71.
58. Hontzeas N, Zoidakis J, Glick BR, Abu-Omar MM. Expression and characterization of 1-aminocyclopropane-1-carboxylate deaminase from the rhizobacterium Pseudomonas putida UW4: a key enzyme in bacterial plant growth promotion. Biochim Biophys Acta Proteins Proteomics. 2004b;1703(1):11–9.
59. Jacobson CB, Pasternak JJ, Glick BR. Partial purification and characterization of 1-amino cyclopropane-1-carboxylate deaminase from the plant growth promoting rhizobacterium *Pseudomonas putida* GR12-2. Can J Microbiol. 1994;40(12):1019–25.
60. Jia Y, Kakuta Y, Sugawara M, Igarashi T, Oki N, Kisaki M, Shoji T, Kanetuna Y, Horita T, Matsui H, Honma M. Synthesis and degradation of 1-aminocyclopropane-1-carboxylic acid by *Penicillium citrinum*. Biosci Biotechnol Biochem. 1999;63(3):542–9.
61. Minami R, Uchiyama K, Murakami T, Kawai J, Mikami K, Yamada T, Yokoi D, Ito H, Matsui H, Honma M. Properties, sequence, and synthesis in Escherichia coli of 1-aminocyclopropane-1-carboxylate deaminase from *Hansenula saturnus*. J Biochem. 1998;123(6):1112–8.
62. Ose T, Fujino A, Yao M, Watanabe N, Honma M, Tanaka I. Reaction intermediate structures of 1-aminocyclopropane-1-carboxylate deaminase: insight into PLP-dependent cyclopropane ring-opening reaction. J Biol Chem. 2003;278(42):41069–76.
63. Stearns JC, Woody OZ, McConkey BJ, Glick BR. Effects of bacterial ACC deaminase on Brassica napus gene expression. Mol Plant-Microbe Interact. 2012;25(5):668–76.

64. Walsh C, Pascal RA Jr, Johnston M, Raines R, Dikshit D, Krantz A, Honma M. Mechanistic studies on the pyridoxal phosphate enzyme 1-aminocyclopropane-l-carboxylate deaminase from *Pseudomonas* sp. Biochemistry. 1981;**20**(26):7509–19.
65. Christen P, Metzler DE. Transaminases. New York: John Wiley and Sons; 1985.
66. Glick BR. Modulation of plant ethylene levels by the bacterial enzyme ACC deaminase. FEMS Microbiol Lett. 2005;251(1):1–7.
67. Glick BR, Penrose DM, Li J. A model for the lowering of plant ethylene concentrations by plant growth-promoting bacteria. J Theor Biol. 1998;190(1):63–8.
68. Penrose DM, Moffatt BA, Glick BR. Determination of 1-aminocycopropane-1-carboxylic acid (ACC) to assess the effects of ACC deaminase-containing bacteria on roots of canola seedlings. Can J Microbiol. 2001;47(1):77–80.
69. Sessitsch A, Coenye T, Sturz AV, Vandamme P, Barka EA, Salles JF, Van Elsas JD, Faure D, Reiter B, Glick BR, Wang-Pruski G, Nowak J. *Burkholderia phytofirmans* sp. nov., a novel plant-associated bacterium with plant-beneficial properties. Int J Syst Evol Microbiol. 2005;55(3):1187–92.
70. Sun Y, Cheng Z, Glick BR. The presence of a 1-aminocyclopropane-1-carboxylate (ACC) deaminase deletion mutation alters the physiology of the endophytic plant growth-promoting bacterium *Burkholderia phytofirmans* PsJN. FEMS Microbiol Lett. 2009;296(1):131–6.
71. Abeles FB, Morgan PW, Salveit ME. Ethylene in plant biology. 2nd ed. San Diego: Academic Press; 1992. p. 414.
72. Frankenberger WT, Arshad M. Microbial synthesis of auxins. In: Arshad M, Frankenberger WT, editors. Phytohormones in soils. New York: Dekker; 1995. p. 35–71.
73. Glick BR. Bacterial ACC deaminase and the alleviation of plant stress. Adv Appl Microbiol. 2004;56:291–312.
74. Jackson MB. Ethylene in root growth and development. In: Mattoo AK, Suttle JC, editors. The plant hormone ethylene. Boca Raton, FL: CRC; 1991. p. 169–81.
75. Abbamondi GR, Tommonaro G, Weyens N, Thijs S, Sillen W, Gkorezis P, Iodine C, Rangel WM, Nicolaus B, Vangronsveld J. Plant growth-promoting effects of rhizospheric and endophytic bacteria associated with different tomato cultivars and new tomato hybrids. Chem Biol Technol Agric. 2016;3:1.
76. Khan AL, Halo BA, Elyassi A, Ali S, Al-Hosni K, Hussain J, Al-Harrasi A, Lee I. Indole acetic acid and ACC deaminase from endophytic bacteria improves the growth of *Solanum lycopersicum*. Electron J Biotechnol. 2016;21:58–64.
77. Onofre-Lemus J, Hernández-Lucas I, Girard L, Caballero-Mellado J. ACC (1-aminocyclopr opane-1-carboxylate) deaminase activity, a widespread trait in *Burkholderia* species, and its growth-promoting effect on tomato plants. Appl Environ Microbiol. 2009;75(20):6581–90.
78. Raweekul W, Wuttitummaporn S, Sodchuen W, Kittiwongwattana C. Plant growth promotion by endophytic bacteria Isolated from rice (*Oryza sativa*). Thammasat Int J Sci Technol. 2016;21(1):6–17.
79. Jasim B, Joseph AA, John CM, Mathew J, Radhakrishnan EK. Isolation and characterization of plant growth promotingendophytic bacteria from the rhizome of *Zingiber officinale*. 3 Biotech. 2014;4:197–204.
80. Lowman JS, Lava-Chavez A, Kim-Dura S, Flinn B, Nowak J, Mei C. Switchgrass field performance on two soils as affected by bacterization of seedlings with *Burkholderia phytofirmans* strain PsJN. Bioenergy Res. 2015;8(1):440–9.
81. Kim-Dura S, Lowman S, Zhang S, Mei C. Growth promotion of switchgrass by bacterial endophyte *Pantoea agglomerans* strain PaKM isolated from seeds. J Pathol Microbiol. 2016;1(2):1007.
82. Gamalero E, Marzachì C, Galetto L, Veratti F, Massa N, Bona E, Novello G, Glick BR, Ali S, Cantamessa S, D'Agostino G, Berta G. An 1-Aminocyclopropane-1-carboxylate (ACC) deaminase-expressing endophyte increases plant resistance to flavescence dorée phytoplasma infection. Plant Biosyst. 2016;(151):331–340.

83. Han Y, Wang R, Yang Z, Zhan Y, Ma Y, Ping S, Zhang L, Lin M, Yan Y. 1-Aminocyclopropane-1-carboxylate deaminase from *Pseudomonas stutzeri* A1501 facilitates the growth of rice in the presence of salt or heavy metals. J Microbiol Biotechnol. 2015;25(7):1119–28.
84. Yaish MW, Antony I, Glick BR. Isolation and characterization of endophytic plant growth-promoting bacteria from date palm tree (*Phoenix dactylifera* L.) and their potential role in salinity tolerance. Antonie Van Leeuwenhoek. 2015;107(6):1519–32.
85. Barnawal D, Bharti N, Tripathi A, Pandey SS, Chanotiya CS, Kalra A. ACC-deaminase-producing endophyte Brachybacterium paraconglomeratum strain SMR20 ameliorates chlorophytum salinity stress via altering phytohormone generation. J Plant Growth Regul. 2016;35(2):553–64.
86. Sziderics AH, Rasche F, Trognitz F, Sessitsch A, Wilhelm E. Bacterial endophytes contribute to abiotic stress adaptation in pepper plants (*Capsicum annuum* L.). Can J Microbiol. 2007;53(11):1195–202.
87. Jung HW, Kim W, Hwang BK. Three pathogen-inducible genes encoding lipid transfer protein from pepper are differentially activated by pathogens, abiotic, and environmental stresses. Plant Cell Environ. 2003;26(6):915–28.
88. Iniguez AL, Dong Y, Carter HD, Ahmer BMM, Stone JM, Triplett EW. Regulation of enteric endophytic bacterial colonization by plant defenses. Mol Plant-Microbe Interact. 2005;18(2):169–78.
89. Ali S, Duan J, Charles TC, Glick BR. A bioinformatics approach to the determination of genes involved in endophytic behavior in *Burkholderia* spp. J Theor Biol. 2014b;343:193–8.
90. Rahemi M, Jamali B. Carnation flower senescence as influenced by nickel, cobalt and silicon. J Biol Environ Sci. 2011;5:147–52.
91. Reid MS, Wu M. Ethylene and flower senescence. Plant Growth Regul. 1992;11(1):37–43.

Index

S.L. Doty (ed.), *Functional Importance of the Plant Microbiome*,
DOI 10.1007/978-3-319-65897-1

O

P

R

S

FSC
www.fsc.org
MIX
Papier | Fördert gute Waldnutzung
FSC® C083411

Kolibri 360 GmbH
Ferdinand-Jühlke-Straße 7
99095 Erfurt, Deutschland
produktsicherheit@kolibri360.de